The Great Northern Coalfield
1700-1900

Illustrated notes on the Durham and Northumberland Coalfield

Frank Atkinson

University Tutorial Press Ltd
9-10 Great Sutton Street, London, E.C.1

All rights reserved. No portion of the book may be reproduced by any process without written permission from the publishers.

© Frank Atkinson 1966, 1968

First published by Durham County Local History Society 1966

First published in this edition 1968

SBN: 7231 0170 1

PRINTED IN GREAT BRITAIN BY UNIVERSITY TUTORIAL PRESS LTD, FOXTON, NEAR CAMBRIDGE

... the Number of Arts and curious Machineries that are used in this Affair of the Coal Business; the sinking of Pits, Winning of Coals, Fire-Engines, Waggon-Ways, the Waggons, their managing of them, the Staithes &c. ...

HENRY BOURNE
History of Newcastle Upon Tyne
1736

In proceeding to notice the methods of obtaining the coal, and the general economy of a mine, we will refer to the collieries of Northumberland and Durham, as exhibiting the most perfect arrangements that have yet been made in this department of mining.

CHARLES TOMLINSON
Cyclopaedia of Useful Arts and Manufactures
1857

Thanks are due to the following publishers for the loan of blocks: David and Charles (*Thomas Newcomen*) (Fig. 14); Methuen (*Coal Mines and Miners*) (Fig. 32); Nelson (*North England*) (Fig. 43); Radio Times Library (Fig. 32); and the Editor of the Northumberland and Durham Architectural and Archaelogical Society (Figs. 11, 16, 21, and 23). For the loan of all other blocks the publishers express their thanks to the Durham County Local History Society. The National Coal Board is thanked for two photographs (Figs. 54 and 55), and the Stephenson Engineering Society for permission to reproduce two illustrations (Figs. 25 and 46 the latter being drawn by Christopher Wright). The drawings for Figs. 5, 6, 22, and 38 are by Roy Varndell and the photographs for Plates 26, 34, 36, 39, 48, 50, 51, 57, and 58 are by the author.

Jacket. PITMEN NEAR CROOK, CO. DURHAM, ABOUT 1890.

Contents

5 Introduction

Preparation
6 Boring
8 Sinking

Mining
11 Methods of mining
13 Hewing
16 Ventilation
21 Lighting
23 Drainage
27 Underground transport
31 Winding
37 Surface treatment
42 Accidents

The Pitmen
44 Housing
47 Clothing
47 Wages

Surface Transport
48 Horse wagonways
52 Steam haulage
55 Railways
57 Staithes
61 Shipping

Local Uses
62 Coking and By-Products
70 Saltmaking
71 Iron and steel and glass

72 Bibliography

75 Index

Introduction

Mining history in the coalfields of Durham and Northumberland could be examined from several viewpoints: of economic factors; of accidents and improving conditions; of the bitter fighting of miners and owners. These historical approaches are well documented, as also are such ancillary subjects as the railway systems. But the casual reader may have difficulty in finding specific descriptions and detailed illustrations of the techniques and equipment of the past two centuries. One hopes that this booklet will make this industry seem more interesting and perhaps easier to understand.

Additionally, to look at the three dimensional objects of history can serve a more positive purpose. We shall see over the next few pages how elaborate some early equipment could be: a surprise and antidote to those who think that only the things of to-day are elaborate and "sophisticated". Considerable technical ability, admittedly much of it empirical, can moreover be recognised. But throughout our period we shall also see traditional methods and equipment often continuing in use, practically without change, alongside greatly improved methods. Would one have expected a horse to be winding coal in 1942, an 1836 steam engine still to be working in 1960, or a beehive coke oven of a type invented in 1750 still to be in use in 1950?

We shall see how the early adoption of a technical improvement might prevent or delay the subsequent acceptance of further improvements. (For example, the early use of wheeled trams to carry the corves from the coal face, in place of sledges, delayed acceptance in the North of wheeled tubs or corves-on-wheels for forty years, in comparison with the Yorkshire and Derbyshire coalfields.) Moreover, technical progress was not always to the advantage of the worker: in backward coalfields such as Tipperary and Kilkenny, where the tramway was unknown in 1842, strong young men "hurried" the coal, and child labour underground was almost unknown. Yet in Durham and Northumberland where tramways were in general use, the putters were said in 1842 to be younger than their predecessors of the days before the invention of the wheeled corf.[1]

Here, then, is another approach to history, and one hopes that the present interest in Industrial Archaeology, the local historians with mud on their boots, and the "new look" of open air Museums will together lead to a generation of historians whose minds will question beyond the printed and manuscript source to the actual objects. Certainly an author writing from within a Museum finds himself thinking of the material objects still existing, from pit lamps to winding engines, and it is from this point of view that the present notes are written.

This booklet has been based wherever possible on contemporary descriptions or illustrations, and an attempt has been made to avoid the more commonly used illustrations. All sources, where practicable, relate specifically to the north-east of England.

For the second half of the nineteenth century one might expect photographs to be available as accurate records, but relatively few are preserved. A good collection of about 1890 to 1900 from South Shields has been reserved for the Regional Museum and a few are used here. Perhaps as a result of this publication others will come to light.

[1] Ashton and Sykes, 1929: 173.

PREPARATION

Boring

Coal would first be found and worked where the seam *outcropped* or appeared at the surface. Later, experience of shallow seams would permit small pits to be sunk the necessary few feet. Working out into the coal from the foot of such a shaft would result, as the roof partially collapsed, in what we know as a "bell-pit".

As the shallow, easily discovered, seams were worked out and deeper seams were sought, the cost of sinking exploratory shafts led to a cheaper method of investigation by borehole. George Sinclair in *The Hydrostaticks* (1672) wrote: "To find Coal, where never any hath hitherto been discovered . . . there are but three wayes. First by sinking, which is most chargeable. . . . There was a second way invented to supply this defect, which is by boaring, with an instrument made of several Rods of Iron . . ." His third method was by ranging or estimating the whereabouts of seams from their observed outcrop and dip.

It seems likely that boring for coal began in England in the very early years of the seventeenth century, and according to Gray in *Chorographia* (1649) boring rods were introduced to the north about 1605 by Huntingdon Beaumont, "an ingenious gentleman" whose family had coal mines in Leicestershire.[1] Certainly by 1618 rods were used for boring at Newcastle, since the Household Books of Naworth Castle record a payment on 2nd July, 1618, for a "sett of boaring rods bought at Newcastell: v li. xvi s. ix d." (£5 16s. 9d.).[2] Nevertheless, there was still some doubt in the mid-seventeenth century as to the merits of boring, and both Sinclair and later Robert Plot (in *Natural History of Staffordshire*, 1686) considered that when the coal lay deep, boring became scarcely less tedious and expensive than sinking trial shafts: "the drawing of the Rodes consuming so much time, in regard it must be frequently done". Yet by 1708 when J.C. published *The Compleat Collier*, techniques must have improved, and the method was an accepted one. J.C. described the process as carried on in the "Great Northern Coalfield" thus: "We have two labourers at a time, at the handle of the bore Rod, and they chop or pounce with their Hands up and down to cut the Stone or Mineral, going round, which of course grinds either of them small, so that finding your Rod to have cut down four or six Inches, they lift up the Rod, either all at once, as there is conveniency for its Lift; or by Joynts fixing the Key which is to keep the Rod from droping down into the hole . . . and taking off the cutting Chissel, puts or screws on the Whimble or Scoop which takes up the cut Stuff be it what happens . . .".

Almost eighty years later John Brand[3] described the tools of the borer then in use around Newcastle, which would seem to have changed little over that period: "(They) are of the simplest kind and consist of iron rods, each between three and four feet long, and about an inch and a half square, with a screw (a bed and what fits into it) at each end, by which they are united, and lengthened out as the hole increases in its depth. The chissel is about eighteen inches long and two and a half broad at the end, and is fixed to the lowest

[1] Smith, R. S., 1957. [2] Galloway, 1898: 152. [3] Brand, 1789: II, 678.

of the rods. The uppermost rod is furnished with an eye, wherein to insert a piece of timber for an handle for two or more workmen to take hold of at once".

Brand adds in a footnote: "After they get down to a certain depth, the rods are wrought by a bracke; a box of wood is inserted first into the ground, to keep the rods in a straight direction. A triangle is erected over the spot where the boring is made, for the sake of drawing up the rods". A hundred and thirty years after *The Compleat Collier* was published and fifty years after John Brand's *Newcastle*, another writer described and illustrated very similar equipment and methods. Fig. 1 is taken from John Holland's *Fossil Fuel* (1835). The springy pole (A on the diagram) enabled the borers to lift the rods high enough to give sufficient impetus to the next blow, for it would impart a springing rhythm. This is almost certainly what John Brand meant when he wrote ". . . the rods are wrought by a bracke". Hughes' *Text Book of Coal-Mining*, published in 1892, mentioned this equipment as currently in use at the end of the century.

Although boring by chisel-bit continued in use and is still used on a small scale to-day, a method of cutting a core of rock was patented in 1805, thus enabling the rocks to be more closely examined. In about 1846 a method of boring with hollow rods, combined with a force-pump, was introduced. The detritus cut by the tool was removed continuously by a circulating current of water and the debris caught in settling troughs for examination.

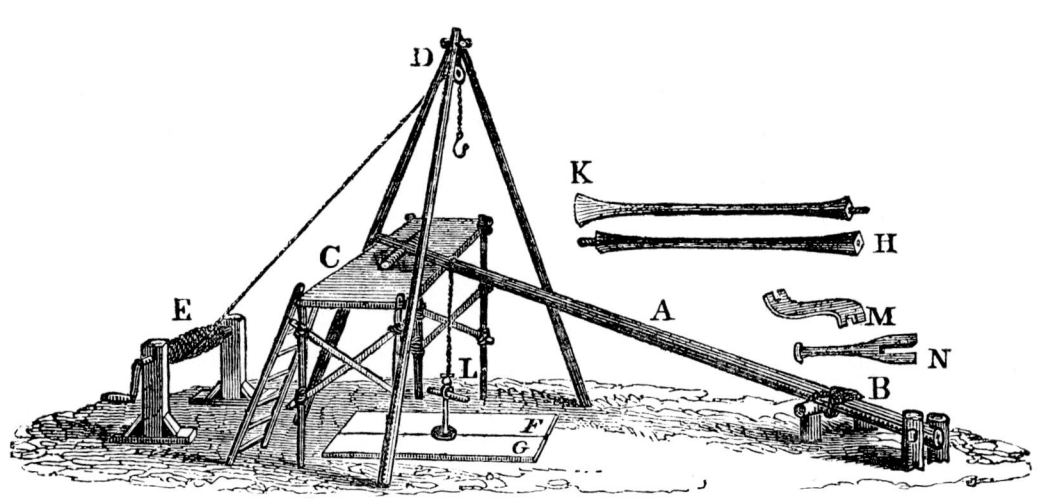

Fig. 1. BORING EQUIPMENT, 1835.

Rods H, K have swollen ends for added strength. The lowest rod is chisel-ended (K), M is a spanner, N an iron fork to grip the rods below the 'swell', to hold the lower rods while one is being screwed off or on. (From Holland's *Fossil Fuel*.)

Sinking

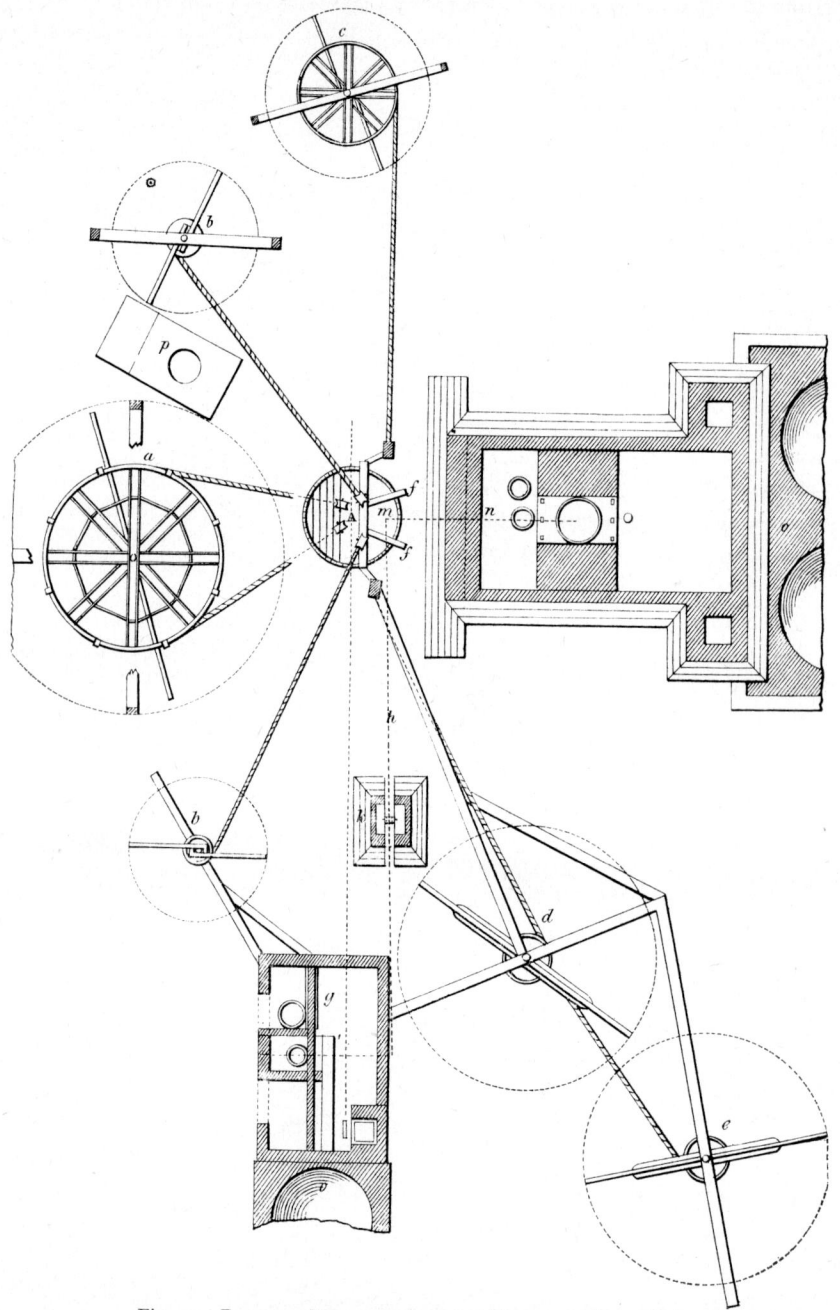

Fig. 2. Plan of Machinery when Sinking a Pit, 1848.
(From Dunn's *Winning and Working Collieries*.)

When coal has been proved underground by boring, the next operation in coal-mining is the sinking of a shaft. We have a description of northern methods from *The Compleat Collier* (1708). A four-sided pit was first cut in the surface soil, but as the sinking proceeded towards the stone it was shaped to an octagon, and the shaft through the stone itself was circular in plan. The sides above the stone were timbered with fir balks and lined with deal boards to prevent falls of earth. When sand was met with it was usual to hold it back by ramming clay between it and this wooden *tubbing*. When the sinkers came to wet strata they sometimes packed undressed sheepskins between the boards and the stone, and sometimes the sides were lined with bricks, behind which spiral channels, known as *garlands*, were constructed to carry to the pit bottom water which would otherwise have forced its way into the shaft. The square shaft at ground level was 'nine quarters' across, or 6 ft 9 in., and if we deduct from this the timbering, it only leaves a diameter of about 6 ft for the finished pit.

The whole work of sinking, as late as the early years of the eighteenth century, was done manually, and the ordinary tools of the sinker would be mattocks, a gavelock (a stout iron bar), a sledge hammer, and several short wedges. Although the use of gunpowder for blasting was introduced to this country early in the seventeenth century by German miners, it was used only in metal mines for many years. The first known record of gunpowder being used in a coal mine is in 1719 and by 1750 it was being employed at Lanchester Moor Colliery.[1] The equipment and methods of shot-firing are described later.

In the early eighteenth century a windlass over the shaft mouth would serve to wind out the excavated material, but by the nineteenth century pits were being sunk larger and deeper, requiring more elaborate equipment. Dunn (1848)[2] tells us the surface requirements for a shaft to be sunk 100 fathoms deep. First, there must be, in addition to the implements to be handled by the sinkers, wrights' and smiths' shops, cottage houses, office, crabs, shear-legs, and winding and pumping engine, ponds, fire-lamps, sinker's lodge, and heap-stead. These are shown in his plate (Fig. 2).

> Two *ground crabs* (b) or horse-turned winches, support the pumps. They are only turned when the pumps require adjustment. The *main crab* (d), backed by the *tail crab* (e) is used to raise and lower the pumps, spears, or other heavy equipment in the engine shaft.
> The *jack gin* (c) is capable of slow movement, and being more controllable is used to lift men during the fixing of pumps or other work in the shaft.
> The *sinking gin* (a) is only a temporary arrangement, giving way eventually to the permanent winding engine.
> The main pumping engine (n) is assisted during sinking by the winding engine (g), to the axle of which is attached a temporary *pumping beam* (h) supported at its pivot by a stone pillar (k).

Dunn next describes the method of ventilation which became necessary when the shaft had reached 10 or 12 fathoms. A wooden partition or *brattice* was built across the shaft and, by means of a small furnace, air was circulated. The brattice was extended down as the shaft was sunk deeper (Fig. 3).

Another improvement described by Dunn as being recently adopted (1848), was the provision of a movable iron hut, which remained at the bottom of the pit. It saved the loss of time occasioned by men ascending and descending, for they retired into this metal

---

[1] Ashton and Sykes, 1929: 15. [2] Dunn, 1848: 43.

hut whilst the shots exploded. One such hut still stands at the water pumping station at New Heseldon, Co. Durham. It was probably used during the sinking of that pumping shaft in 1879.

On completion of the shaft as far as the coal, a sump was further excavated to provide *standage* for the water. Indeed additional reservoir space was often cut out of the rock to provide for a few days' or even weeks' accumulation of water, should the pumps fail or the water unexpectedly flood.

Bratticing was then permanently fixed, to provide air circulation and other needs. It must be remembered that not until 1862 did it become compulsory to provide a second shaft and most deep collieries made multiple use of the one entrance. In fact a nineteenth century glossary[1] states that the word *shaft* can mean part of a *pit*, separated off by a brattice, and quotes: "a pit divided by a brattice into two shafts, *viz.*: a coal and an engine or water shaft".

[1] Greenwell, 1888.

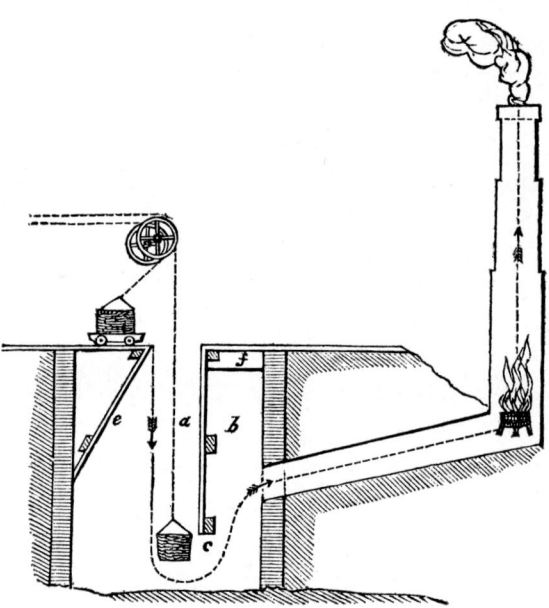

Fig. 3. VENTILATING A SHAFT WHILST SINKING IT, 1848.
(From Dunn's *Winning and Working Collieries.*)

MINING

Methods of Mining

Two main methods of excavating the coal have been in use in this country and were described in the early nineteenth century as *long* or broad and *short* or narrow methods. The former, by which nearly all the coal intended to be got is dug out at once, has been commonly in use in Yorkshire and elsewhere for two centuries or more, where the strata immediately above and below the coal are hard and firm. The latter method, commonly termed *bord* (or board) and *pillar*, or *pillar and stall*, was in general use in the Durham and Northumberland coalfield, and is still practised, though over rapidly diminishing areas. It was believed to be the only safe method of working a tender or fairly soft coal which also had unsafe floor and roof, so that large portions of the seam had to be left to support the roof (Fig. 4).

Such pillar and stall working has a long history. An example which cannot be later than the fifteenth century is now being extracted near Butterknowle,[1] Bishop Auckland, and although irregular, shows this principle at work. Seventeenth century examples are known from documentary evidence,[2] and at the beginning of the eighteenth century *The Compleat Collier* described the working places as being 3 yds in breadth, and added that between one bord and the next a pillar, 4 yds across, was left standing. Thus less than half the coal was removed from the mine.

In succeeding years dimensions of bords and pillars increased: in the mid-eighteenth century a colliery near Berwick had bords about 10 ft wide and pillars 18 ft long and 10 ft wide, and by 1835 typical examples were given by John Holland as having bords 4 or 5 yds wide, leaving pillars 20 yds by 9, or sometimes about 9 yds square.

Until probably the 1730s the pillars were left in their entirety, but around this time began a deliberate 'robbing' of the pillars either by breaking away their ends, or by *jenkins* cut through them. But in the deeper workings around Newcastle the roof pressure and risk of explosion made this second working a dangerous undertaking. It was not until 1795 that a partial solution was found by Thomas Barnes, described as *panel working*. His first experiments were in 'fiery collieries below Bridge',[3] and amounted to dividing the colliery into sections, each of about 10 to 20 acres, round which were built *biggins* or walls of stone about 40 yds thick. A variation to this method lay in leaving complete barriers of unworked coal up to 50 yds in thickness. Within each panel alternate pillars could be removed, and the whole finally sealed off to prevent the escape of gas, when completed. Under such circumstances the methodical approach of the nineteenth century led to a consideration of the optimum dimensions of coal pillars, at various depths, and Matthias Dunn published in 1848 a table showing dimensions for safe working.

---

[1] Four fifteenth century bronze cooking vessels were discovered in old workings in 1964 and are now in The Bowes Museum.
[2] Nef, 1932: I, 365.
[3] *i.e.* downstream from Newcastle bridge.

Finally, may we quote one example of Durham practice in the early twentieth century. In 1964, John Kell, then aged seventy-seven, described his working life in Leasingthorne Colliery, near Bishop Auckland. He worked most of his time in the Brockwell seam (4 ft thick). The method of working was "*bords and walls*: (that's what the miner calls it: some people call it pillar working)". First *the winnin* was cut: a passage 3 or 4 yds wide and which may be extended up to say two miles or more. (The Brockwell

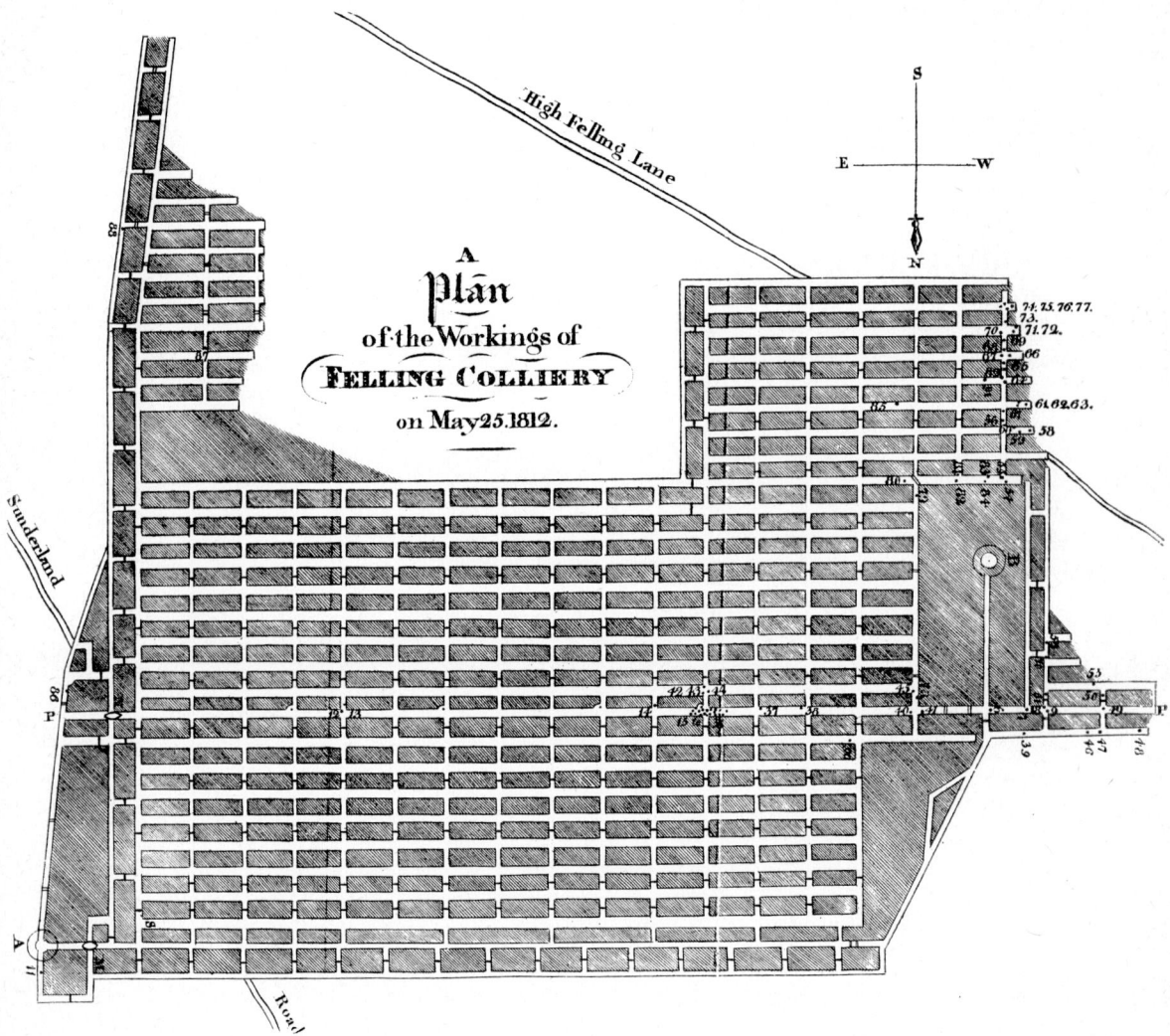

Fig. 4. PLAN OF A COLLIERY SHOWING PILLAR AND STALL WORKINGS, 1812.

Plan made after explosion of 1812. *A* is down-cast, working shaft and *B* up-cast shaft. *Crane* for transferring corves was near centre of plan (see cluster of numbered dots, indicating positions of men and boys killed). (From Hodgson: *Funeral sermon*.)

seam had a good hard bottom and a good hard top.) Then cuts were made into the coal on each side, 5 yds wide, and after 30 yds you turned left and right, to leave pillars 30 feet square. One way was the *bord*; and at right angles was the wall. "When you are working bord the coal is facing you; when you are working wall, the coal is end on. So it is easier to work bord." John Kell compared this with sawing a plank of wood: it is easier to saw it across than down its length. "But sometimes you worked it *cross-cut*" (*i.e.* diagonally) when there was little difference in the two directions.

Hewing

The essential process of hewing or getting coal is one that has been carried on far from the casual traveller, and such contemporary descriptions of early mining methods as exist rarely give detailed evidence of the actual techniques. An early description is that given by John Brand in 1789: "The hewer first digs as far as he can into the bottom of the stratum; then he nooks or corners off the part measured out, and afterwards the great coals come away by a wedge or mallet". John Holland in 1835 wrote: "The men in the boards or stalls first undermine the mass, by hewing out a portion next the *thill* or floor of the mine. This operation is called *kerving* or *holing under*; and it is the object of the workmen to waste as little of the coal as possible . . .; in some instances it has a sole or footing of inferior value, which is the more cheerfully sacrificed. In the next place, deep vertical grooves are cut on each side of the mass intended to be brought away, so as to define its size: it is then broken down, either by means of wedges or by the force of gunpowder; the former is mostly the method adopted in Yorkshire and various other places; the latter generally prevails in the deep mines of the more northern counties. In blasting, a hole is made about a yard in depth, and the shot inserted near the roof or top of the working; and as the effect of the explosion is much more considerable when it can be made to act mainly in the line of one of the vertical back fissures, the collier . . . acquires such an exact perception of the nature of the substance, that he can generally pronounce when his chisel reaches one of these polished faces".

In 1849, according to Greenwell, it was still common for a hewer to work his own board or wall, but double-working was coming into practice, whereby two hewers worked together. At that time an addition of 2d. per yard to the yard price, or 3d. or 4d. per score to the score price was frequently made for the inconvenience supposed to be attached to this manner of working. A document of 1802 shows additional payment of 3d. per score for 'Doulle Boreded' working (Fig. 62). By the early twentieth century it had become common practice to work in pairs, as *marrers*, and John Kell (see above) worked in a group of six marrers. Two brothers, two cousins and two others worked together as marrers for many years. They worked in pairs over three shifts: foreshift century, the powder was placed beforehand in a brown paper tube, to form a cartridge, and John Kell remembers his father, seated before the fire, preparing his cartridges for work next day. The cartridges were about an inch in diameter and from 4 to 9 or 10 ins. long. The *pricker*, a rod of copper about ¼ in. in diameter, tapering to a point and about 3 ft long, was left in the hole during tamping, and on its withdrawal a small channel was left down the side of the hole, as far as the cartridge. Iron prickers had resulted in a number of accidents due to sparks being struck from stone or pyrites, and by the Mines Act of 1872 iron was forbidden for this purpose. Into the channel left by the pricker a piece of straw filled with a priming of powder (the *kitty*) was inserted to reach as far as the charge. Finally a match of greased twine, or a small piece of candle end, was attached. This would burn long enough to give the miner time to retire to safety.[1]

Ventilation

Early mines, shallow and of small extent, would not require assisted ventilation. A certain amount of natural air circulation would be created by changes in barometric pressure, for a rise would drive air into the mine and a drop would draw it out, but as mine workings extended a ventilating shaft had to be sunk at some distance from the winding shaft. George Sinclair writing in 1672 about *choke damp*, or carbon dioxide, said: "The cause seems to be, that the Air underground, wants communication with the Air above ground, because it is found, that by giving more communication, the evil is cured. . . . Some are of opinion that this defect might be supplied by the blowing of Bellows, from above ground, through a Stroop of leather, which must run along to the end of the Level. . . . But I have not yet heard, that it hath been made practicable." On occasion the expense of a vent-pit could be avoided by driving a *vent-head* or tunnel into another working pit or an old abandoned shaft, thus obtaining a circulation of fresh air.

During the seventeenth century artificial devices were produced to force out or draw off the stagnant air, and Plot mentions a method "which they use about Chedle", by which "they let down (fire) in an Iron cradle, they call their lamp, into the shaft", so causing air circulation. At some places in the early eighteenth century instead of suspending such a fire basket in the shaft, a furnace was constructed at the foot of the shaft, and a visitor to a pit at Long Benton in 1749 wrote: "A large lamp stands at the Bottom of this shaft, which they keep in continual blaze for the Convenience of Air, etc., which makes the Shaft as bad to Ride as a Kitchen Chimney".[2]

Long before this, in 1665, in the first volume of the Philosophical Transactions of the newly formed Royal Society, Sir Robert Moray described how, at Liège, he had seen mines worked without the expense of additional air-shafts. A small tower was built near the winding shaft, with a basket of fire suspended in it and from this a wooden pipe

[1] Greenwell, 1888. [2] Ashton and Sykes, 1929: 47.

Fig. 7. Hebburn Colliery, 1844, with Ventilating Shaft *left*, and Screens *right*. (From Hair's *Coal Mines*.) *Corves* are being raised by winding engine (left).

was carried into the workings, so drawing out the foul air. He wrote of the square wooden tube or pipe: "the Joints and Chinks are so stopt with Parchment pasted or glewed upon them, that the Air can no where get in to the Pipe but at the end: And this Pipe is still lengthened, as the Adit or Pit advanceth, by fitting the new Pipes so, as one end is alwaies thrust into the other, and the Joints and Chinks still carefully cemented and stopt as before".

We have no reference to anyone taking notice of this useful technique, until Sir John Clerk of Penicuik, journeying south from Edinburgh in 1724 to see how mining was carried on around Newcastle, saw some of Alderman Richard Ridley's collieries, probably at Byker, and learned from the 'Engineer' the "methode which he used to drain ill aire out of Alderman Ridley's works and likeways to carry a mine a great way without the expense of letting down a shaft".[1] Mr Denald, the Swedish 'Engineer' explained his method and Clerk jotted down a sketch in his Journal, which is shown here (Fig. 11), "At the mouth of the shaft A he made the furnace B & at the aire hole C he fixed the pipe D of timber which was let down the shaft A and by degrees convoyed to the end of the mine or wall face at E".

A similar method was used on the Wear in 1760, when 2 ins. diameter wooden[2] pipes were carried from a surface furnace to remoter workings, and as late as 1848 Matthias Dunn wrote that "until lately it was quite common to see collieries dependent for their ventilation upon a wooden box of 12 or 14 ins. square, sometimes led to the engine fire as a substitute for a proper furnace, and sometimes operated upon by a small revolving fanner, worked by a steam engine".

Other methods were resorted to during the eighteenth century, for the problem, particularly in the Tyne coalfield, was not merely one of "ill aire", but of explosive gases which made it almost impossible to work by the open lights of the time, and a description of 1769 tells of two ventilators worked at Walker Colliery with a machine, "by the help of the fire-engine" (a Newcomen steam engine).[3] Nevertheless, ventilation by fire long continued to be used (Fig. 10) and a technique of circulating fresh air around all the mine workings, known as 'coursing the air' became the practice at collieries on the Tyne and Wear after about 1760 (Fig. 9). This coursing was found necessary because gas tended to accumulate in old workings and a fall of roof or a change in atmospheric pressure might set up a movement of this explosive gas from such an unventilated part of the pit to the working face. By use of *stoppings*, or partitions of brick or wood, the air was driven up and down the course of many workings until it came to the upcast shaft (see also Fig. 4). Where the coal had to be brought across or into such an airway, a wooden door was constructed, and it was the duty of a new class of labour, the child *trappers*, to stay by such a door to open and shut it as required.

Above ground, the fumes from the upcast shaft were assisted upwards and dispersed by a brick or stone chimney and the air intake was assisted by a wind vane atop another funnel. These latter, once a typical sight of the early and mid-nineteenth century Newcastle and Durham coalfield, have all gone, but Fig. 7 shows such a scene around 1840, and Fig. 8 whilst perhaps not so accurate in scale, is of about the same date. During the

[1] Atkinson, F., 1965: 432.   [2] Ashton and Sykes, 1929: 48.   [3] Wallis, 1769: I. 128.

Fig. 8. A WALLSEND COLLIERY, 1835, WITH WINDING ENGINE *centre* AND VENTILATING SHAFT *right*.

A is brick ventilation funnel adjacent to upcast shaft, *D* is head-gear over drawing pit, *E* the engine house, and *F* counterpoise of heavy chain to assist the winding engine. (From Holland's *Fossil Fuel*.)

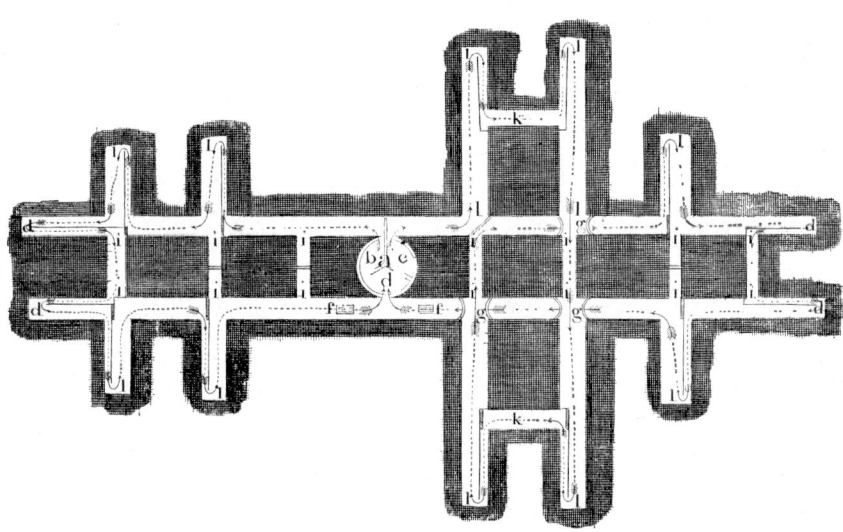

Fig. 9. DIAGRAM OF UNDERGROUND AIR CIRCULATION, 1814. THE PIT SHAFT IS AT *a* AND TWO FURNACES AT *ff*.

The pit *a* is subdivided by vertical *brattices* (wood partitions) into two downcast shafts *b* and *c*, and one upcast *d*. Two furnaces are at *ff* and *ggg* are arches where air currents cross. (From Buddle's *Report* of 1814.)

19

last decades of the nineteenth century mechanical ventilation became practicable and various designs of steam-driven fans were constructed. A Waddle fan, 30 ft in diameter, still stands at Ryhope Colliery, where it was probably installed about 1900. This was driven by a horizontal steam engine at about 100 r.p.m. and exhausted the upcast air, flinging it centrifugally to the atmosphere.

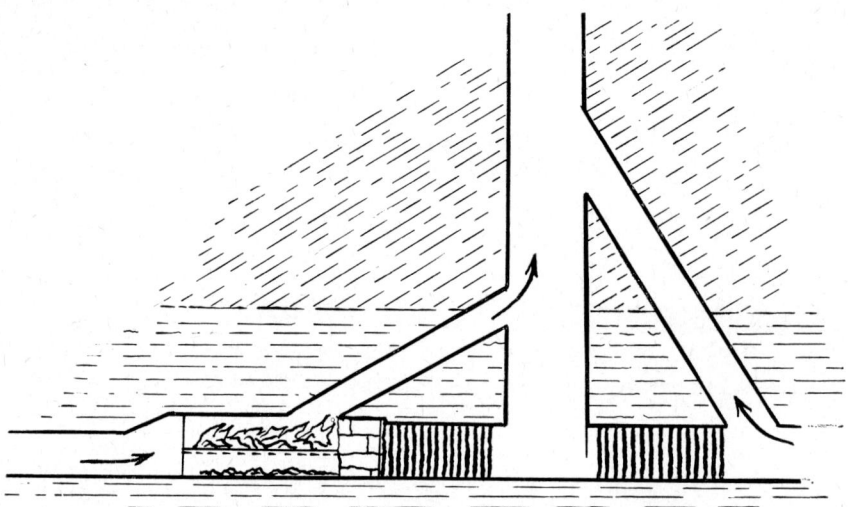

Fig. 10. Vertical Section of the Pit Bottom showing a Ventilating Furnace; Mid-nineteenth Century.

(From Tomlinson's *Cyclopaedia*.)

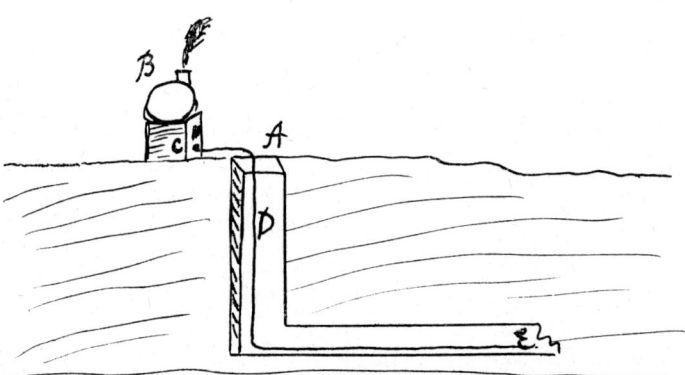

Fig. 11. Ventilation by a Furnace above Ground in 1724.

(Sketch from Sir John Clerk's *Journal*.)

Lighting

Candles were in use in mines throughout the country for several centuries but by the eighteenth century the 'fiery' mines of the Tyne were such that naked flames could not be used and work was sometimes carried out by the feeble light of phosphorus and putrescent fish. Between 1740 and 1750 Carlisle Spedding, of Whitehaven, invented a flint and steel machine, which consisted of a wheel or disc of steel turned at speed by a series of brass gears which were driven by a small handle. The machine was strapped on to a boy, who turned the handle and held a piece of flint against the steel wheel (Fig. 12). The resultant shower of sparks provided a certain amount of light and was believed (though incorrectly) not to ignite gas. According to Dunn *steel mills* (as they were called) were first used in the North East at Fatfield Colliery, on the Wear, after an explosion in 1763. They afforded, he wrote, a glimmering light sufficient to enable a group of five or six miners to carry on their work.

John Buddle in 1813 described how the condition of the sparks from a steel mill could be used to determine the presence of inflammable gas: the sparks increased in size and luminosity, assuming almost a liquid appearance as the percentage of gas rose. Buddle's paper on *The Ventilation of Mines* took the form of the first Report of the Sunderland Society for preventing accidents in coal mines: a Society which had been formed in 1813 following a series of disastrous explosions, including one in 1812 at Felling Colliery when ninety-two lives were lost.

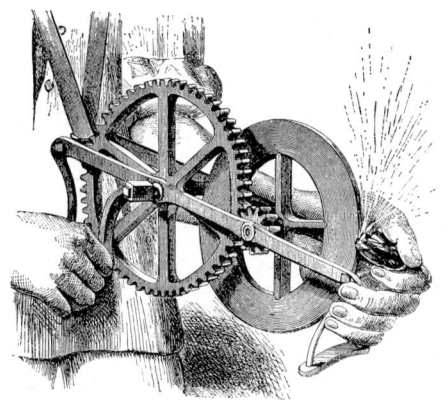

Fig. 12. A Steel Mill in use.
From J. H. Pepper, *Playbook of Metals*, 1861.

This Society invited Sir Humphrey Davy to investigate the problems of lighting mines and in 1815 he produced his safety lamp. Earlier, in 1813, Dr William Clanny, of Sunderland, had produced a complex 'steam safety lamp'. Practically coincidentally with Davy's announcement, George Stephenson produced a similar lamp which long held favour in parts of the North East, where it was known as a 'Geordie Lamp'. Despite the obvious advantages of using the Davy or the Geordie lamp, naked lights continued to be used in many pits. In 1852, for example, the use of candles resulted in an explosion at Seaham Colliery when five men and a ten-year-old boy were killed. Yet naked lights continued well into the present century at 'safe' pits, and many miners still at work to-day began in their youth by the light of *midgie lamps*. (Figs. 13 and 37.)

Improvements in safety lamps included the use of glass for better illumination, but electric battery-operated lamps did not come into general use until about 1910.

Fig. 13. Miners' Lamps. *a*, Mid-Nineteenth Century Davy Lamp; *b*, Late Nineteenth Century Safety Lamp; *c*, 'Midgy' Lamp. The Overall Height of the Centre Lamp (excluding hook) is 10 ins. In the collections of the Northern Regional Open Air Museum.

Drainage

As early as the fifteenth century water made the working of some coal pits difficult and there are documentary references to 'watergates' being cut for drainage. These were probably little more than surface trenches, and an agreement of 1407 entered into by the Sub Prior of Durham, for the lands of Hett, to cut a 'subterranean drain' (*aquaeductum sive trencheam subterraneam*),[1] may equally refer to a drift or trench. In later years drainage channels became extensive and correspondingly costly. By the seventeenth century there were several long water levels in the north-east and one example, the Delaval Drift, began on the Benwell Estate and continued to Whorlton, Newbiggin, and West Kenton.[2] Such levels were cut solely by pick and made as narrow as possible. Greenwell mentions instances of levels sometimes in stone, sometimes in coal, "the width of which does not exceed 18 in.... It is truly a wonder how the work has been performed, the sides of the drift being as smooth and straight as though they had been chiselled".

Coal at depths impossible to drain by levels or adits could only be worked after the water had been lifted out and bailing-out with buckets was a practicable method under certain conditions. As late as the mid-nineteenth century buckets, hauled up the shaft and tipped by a 'trip' mechanism, were still in use. Such a *tub* was described by Matthias Dunn in *Winning and Working Collieries* (1848), as being "strongly ironed and contains about 80 or 100 gallons of water, works upon swivels for emptying, and slots and un-slots in a peculiarly ready manner" (Fig. 18). More than a century earlier Sir John Clerk in 1724 saw similar buckets in use near Newcastle: "... upon slipping off a small sneck or bolt..., the upper end of the Bucket turn'd down by its owne weight & the water was ... convoyed to the level" (Fig. 16). This level to which he referred was an adit designed to carry the water to the River Tyne, obviating water being lifted to the mouth of the shaft.[3]

Yet although buckets, raised by horsepower, were in use in the early eighteenth century and although even more surprisingly they were still in use in the mid-nineteenth century, better and more continuous methods had long existed. In the metal mines of Germany, for example, *rag and chain* pumps were in use in the sixteenth century and Agricola shows several illustrations at work driven variously by man-power, water, and horses (Fig. 15). Pumps are also illustrated, driven similarly. Such methods are known to have existed in England at a yet earlier date, for in 1486 the monks of Finchale spent £9 15s. 0d. on their pump and bought horses to work it and for some years subsequently, in connection with their coal accounts, mention is made of the expenses *de la pompe*.[4]

A seventeenth-century writer, Stephen Primatt, discussing the collieries of the Tyne and Wear says: "In most collieries in the North they make use of Chain-pumps, and do force the same either by Horse Wheels, Tread Wheels, or by Water Wheels; and this they find the surest way for drawing their Water, although the charge of such Wheels for Timber, Leathers, Chains, Pump, and other such materials is very great, besides the great charge of men and horses they are daily at". It seems clear that although the suction pump

[1] Surtees, 1823: III, 287. [2] Edington, 1813: 118. [3] Atkinson, F., 1965: 433.
[4] Galloway, 1898: 71.

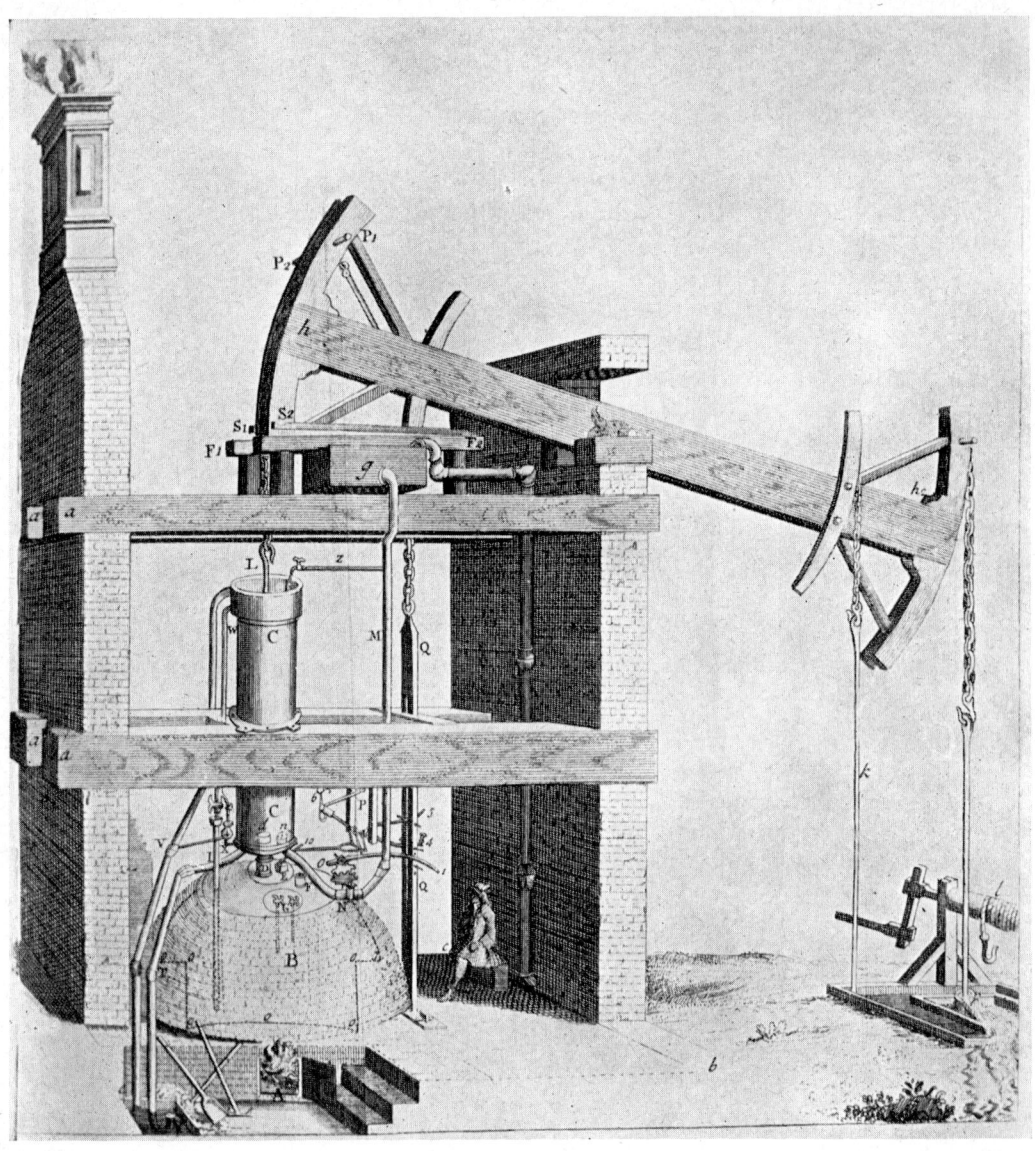

Fig. 14. Newcomen's 'Atmospheric' Engine based on a 1717 Engraving.

From Desagulier's *Experimental Philosophy*, 1744; based on a 1717 engraving of an engine at Griff, Worcestershire.

Fig. 15. A Sixteenth Century 'Rag and Chain' Pump in Germany.

Woodcut from *De Re Metallica* by Agricola, 1556. Round pieces of metal or leather attached to endless chains, were pulled up inside wooden pipes, to lift water.

continued in use in some places in the seventeenth century, the rag-and-chain pump was considered superior; indeed Sir Ralph Delavel, when visited by Lord Guildford in 1676, gave it as his opinion "that chain pumps were the best engine for they draw constant and even".[1] Another description of almost the same date comes from George Sinclair's *Hydrostaticks*, where he details an elaborate system of water-powered chain-pumps built at Ravensworth by Sir Thomas Liddell. The considerable fall of land clearly permitted a waterwheel to be erected high above the ground and the outflow from this powered a lower wheel and a third still lower one. Each wheel operated a chain pump and water was raised by these pumps in three stages.

In 1709 the Earl of Mar sent his colliery manager from Clackmannanshire to inspect machinery in the Newcastle area. From his report (now unfortunately lost, but briefly

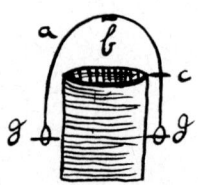

Fig. 16. A Bucket with Quick Release for Raising Water, 1724.

A bucket raised by horse-power. (From Sir John Clerk's *Journal*.)

Fig. 17. The Earliest known Representation of a Steam Engine, on Map of 1715.

This 'Fire Engine' is clearly one of Newcomen's (cf. Fig. 14) and shows chimney, boiler, beam, and pit-shaft. (From a map of the Tanfield Estate.)

Fig. 18. A Bucket for Raising Water, 1848.

(cf. Fig. 16).
(From Dunn's *Winning and Working Collieries*.)

mentioned by Bald in *Coal Trade of Scotland*) it appears that the machines he saw were water-wheels and horse-engines, with chain pumps. The pits were mostly 20 to 30 fathoms deep, though a few reached 60 fathoms.

The first major improvement in drainage methods came in the early eighteenth century when Newcomen perfected his 'Fire Engine' (Fig. 14). There is some doubt as to when and where the first such engine came into the north-east. That there was an engine at Tanfield Lea Colliery in 1715 seems evident from a map of that date showing an unmistakable sketch marked *Engine* (Fig. 17). Two others apparently built not much later were at Oxclose Colliery, Washington Fell, and at Byker Colliery, just north of the Tyne, in Northumberland. The Newcomen engine, operating by atmospheric pressure as steam was condensed in the cylinder, could for many years only be used for a simple vertical movement. Nevertheless, this movement was admirably suited to driving a suction pump and hence this pump once more came into its own, replacing the horse and water-powered chain pumps.

[1] North, 1742, cited by Galloway, 1898: 162.

These steam engines continued to be adopted in the northern coalfield and in 1724 "The Proprietors of the Invention for raising Water by Fire" appointed a Mr John Potter of Chester-le-Street to act as their local agent.[1] Yet little notable improvement took place until 1769 when James Watt separated the condenser from the cylinder and produced a steam, as distinct from an atmospheric, engine. In 1780 the problem of obtaining direct rotary power from the steam engine was solved; by 1782 Watt applied steam alternately on both sides of the piston, giving double the power from the same size of cylinder; in 1784 he patented the parallel action.[2]

Underground Transport

From the hewers the coal was, in the eighteenth century, transported in wickerwork baskets to the foot of the shaft. Sir John Clerk in 1724 wrote: "The coals are drawen from the shaft in what they call corfs made of wands and capable to contain between 2 and 3 Scots loads. The coaliers that work belowe are payed so much a corfe and each of them sends up their owne marked by a Tailey or counter stick. These corfs are brought to the bottom of the shaft on a little slyp shod with iron, and when above ground, they are likeways drawen off on a slyp by a horse used for that purpose".

These baskets, made of hazel wickerwork, are also described in 1789 by John Brand,[3] and illustrated in the 1840s by T. H. Hair (see Figs. 7 and 19). According to Greenwell[4] they had practically ceased to be used by 1888. In the eighteenth century the corf held 4 or 5 cwt.

In the underground conveyance of coal—probably the most arduous of all colliery work—lay a great part of the expense to the coal owner. "The more and a further a pit is wrought . . . the dearer she lies in the charge of barrowmen", wrote J.C. in *The Compleat Collier* (1708). Horses began to be used on the underground roads in the first half of the eighteenth century[5] and when, in 1765, M. Gabriel Jars visited the Walker Colliery he noted that horses were taken down into the mine. He also observed wooden railways, resembling those on the surface, on which four-wheeled rolleys were used to carry the corves. Where there were no railways, young boys used little sleds to draw the corves to the horse gate. This is confirmed by John Brand, writing a few years later, when he explained that a *tram* is a kind of sledge on four wheels. "Barrow-men or coal-putters are a sort of labourers who fill the corves, that are set empty upon sledges or trams (Figs. 21 and 22) with the coals wrought by the hewer, and then draw them, one pulling before and another putting, as they call it, *i.e.* thrusting behind, along the barrow-way to the

[1] Rolt, 1963: 75. [2] Law, 1965: 14. [3] Brand, 1789: II, 681.
[4] Greenwell, 1888: *vide corf*. [5] Ashton and Sykes, 1929: 62.

Fig. 19. An Underground Crane for Transferring Corves, 1844 (see Fig. 20). (From Hair's *Coal Mines*.)

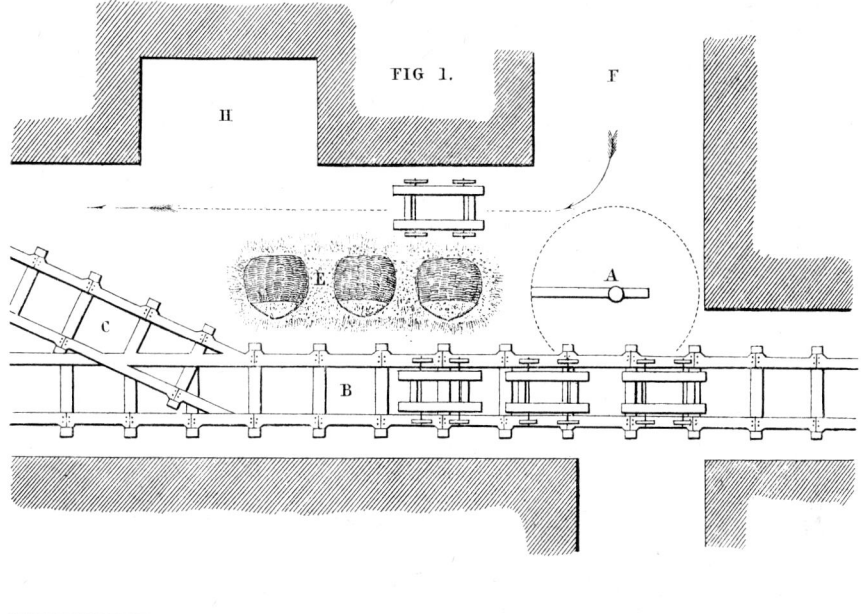

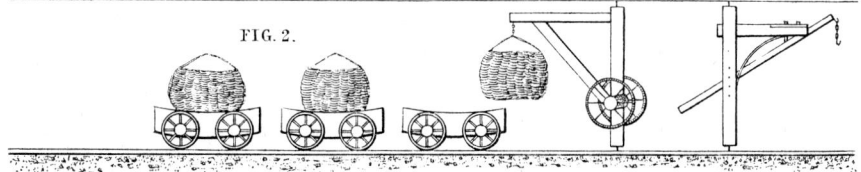

Fig. 20. An Underground Crane for Transferring Corves, 1848 (see Fig. 19).
Crane near pit-bottom used to transfer loaded corves from trams to rolley.
(From Dunn's *Winning and Working Collieries*.)

Fig. 21. A Corf, 1724.
Corf made of hazel.
(After Sir John Clerk.)

Fig. 22. A Tram or Wheeled Sledge, 1768.
To transport corves, at Newcastle.
(After Morand.)

pit-shaft. . . . In low seams (the trams) are pulled by two small cords called *soams* by a boy before, and pushed on at the same time by another boy from behind."[1] For the mid-nineteenth century Greenwell records this same practice, adding that the little boy pulling by the soams was called a *foal* and the one who pushed, aged sixteen or seventeen years, was called the *headsman*.

Early in the nineteenth century we have a description of Felling Colliery, near Gateshead. An explosion there in 1812 killed ninety-two, and a published account[2] includes references to current methods of working (Fig. 4). Coals from the *sheths* or working bords were conveyed in wicker corves on trams to the *crane*. *Barrow-men*, *putters*, and *foals* were employed at this work. The crane lifted the loaded corves off the trams on to wagons or *rolleys* (Figs. 19 and 20), which differed little from the trams except in being larger and stronger. From the crane, four rolleys, each carrying two corves, were taken to the crane bord by way of an inclined plane. On this the empty rolleys were drawn up by the descent of the loaded ones, both sets being attached to a chain passing over an horizontal wheel, controlled by a brake in the charge of a *brakeman*. From the foot of the inclined plane the rolleys were taken to the bottom of the shaft. This inclined plane was reckoned to save the daily expense of at least thirteen horses. At the shaft the corves were lifted off the rolleys by the winding engine and *drawn to bank*; *i.e.* raised up the shaft to the surface (see Fig. 7).

In 1797 John Curr, of Sheffield, published his ideas for a wood-and-iron wheeled truck, to run on cast iron rails. Whilst introducing his idea, he first described recent and current methods: "The prevailing practice till of late, in the working of collieries of Newcastle-Upon-Tyne and Sunderland, was to draw a single corf on a sled from the workings to the shaft of the pit, which as these workings were extended, and the prices and maintenance of horses enormously encreased, became an intolerable burthen to the proprietors of such works; therefore the viewers or superintendents of collieries, have . . . introduced wooden rails, or waggonways underground, for that purpose (or what is generally distinguished by the name of Newcastle-roads) and fixed a frame upon wheels capable of receiving two or three of their basket corves . . . drawn by one horse" (*i.e.* a rolley).

Although Curr's idea of wheeled trucks was adopted in the Yorkshire and Derbyshire coalfield, it curiously was not favoured in the northern area, where corves continued to be pulled on trams and rolleys. It was not until almost 1840[3] that square *tubs* fitted with wheels and containing 6 to 8 cwt. each came into general use here. These were commonly made of wood, and Dunn tells us that they were soon found to be a decided improvement upon baskets.

When first used, the tubs were run on to the rolleys, at the point where previously the crane had transferred the corves from tram to rolley.[4] The rolleys were constructed with transverse dishplates, to keep the tubs in their places on their passage to the shaft. The crane was no longer required, and the place where the tubs were put on the rolleys was called a *flat*. Naturally this complicated arrangement was soon discontinued, and the tubs were drawn by horses along the rolley-ways, without the intervention of rolleys.

[1] Brand, 1789: II.   [2] Hodgson, 1813: II.   [3] Dunn, 1844: 61.
[4] Greenwell, 1888: *vide Crane*.

The development in the northern coalfield may thus be summarised:

Early eighteenth century: corves dragged on sledges from coal face to shaft.
Mid-eighteenth century: corves dragged a little distance, then pulled on trams by horse or putters.
Late eighteenth century: as above, but then corves transferred by crane from tram to rolley (which carried 2 or 3 corves).
c. 1840: wheeled tubs began to be used, being transferred to rolleys for the last (horse-drawn) part of the journey to the shaft.
c. 1860: wheeled tubs brought on their own wheels direct from coal face to shaft, by barrow-man and then by horse. The tub was then carried up the shaft, and tipped above ground.

About 1841[1] engine power was introduced underground, replacing horses where the layout, distance, and traffic justified. Nevertheless, horses continue to be used to the present at many collieries, but are being rapidly reduced in number year by year.

Winding

To wind coal simply by hand windlass would be too slow for large-scale production, and we know the horse was so employed by the sixteenth century. As with so many hand processes, the first attempt at mechanisation simply took the form of applying power to the hand-machine: hence the 'cog and rung' gin. Since the hand-turned windlass hung over the shaft, so did the horse-turned one, and this compelled the horse-track to be constructed around both shaft and windlass. When Sir John Clerk visited Newcastle in 1724, he saw and sketched one (Fig. 23), and wrote: "this Machine is the only one made use of at Newcastle both for drawing up coals and water except the Fire Engine" (*i.e.* Newcomen's steam engine).

A useful watercolour drawing of a gin was made by John Bewick (1760-95), younger brother of Thomas, and is inscribed "Pit at Eltringham": presumably his father's small landsale colliery (Fig. 27). A more diagrammatic but clear illustration was given by that north country mathematician William Emerson in *Principles of Mechanics* (1758) (Fig. 24).

Writing in the *London Magazine* for January, 1765, E. Sarrab, of Chester-le-Street, said: "There are two gins made use of for drawing the coals . . ., a cog-gin and a whim gin. . . . The latter is but of late construction, but esteemed now much superior to the cog-gin . . .". This new development, the whim-gin, seems to have come into use in the Newcastle area

[1] Greenwell, 1888: 36.

probably in the second quarter of the eighteenth century.[1] In this new design the rope-drum was pivoted vertically and the rope brought away from the shaft by pulleys. Thus for the first time in the development of winding machinery the actual winding operation was removed from the pit mouth (Figs. 28 and 29).

Fig. 23. A Cog and Rung Gin, 1724, for Winding from a Shaft by means of a Horse.

The horse-track was round the gears *and* the pit shaft. (From Sir John Clerk's *Journal*.)

"**Description of the Engine**

A is a cylander upon which the rop turns which draws from the pit or shaft F.

B is a horizontal wheel with cogs which turns the cylinder by the spinnel C.

D is the pole fixt to the Axis of the Horizontal wheel, to which the horse is tied.

E is a circle described by points which the horse makes in turning about the Horizontal wheel, under the great beams or stays of the Engine marked G. & H."

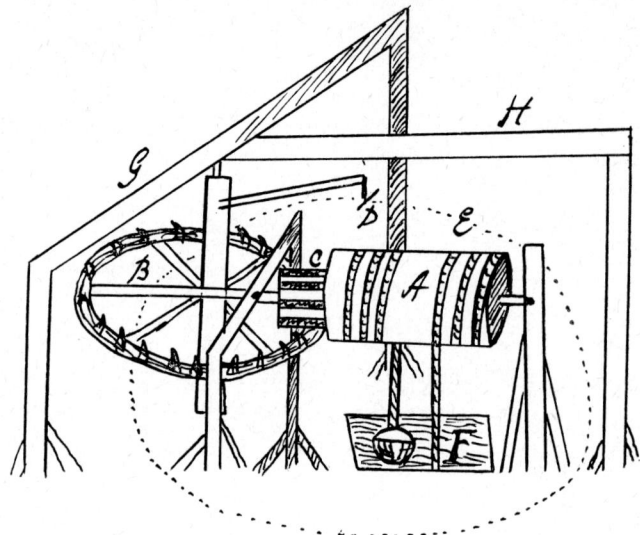

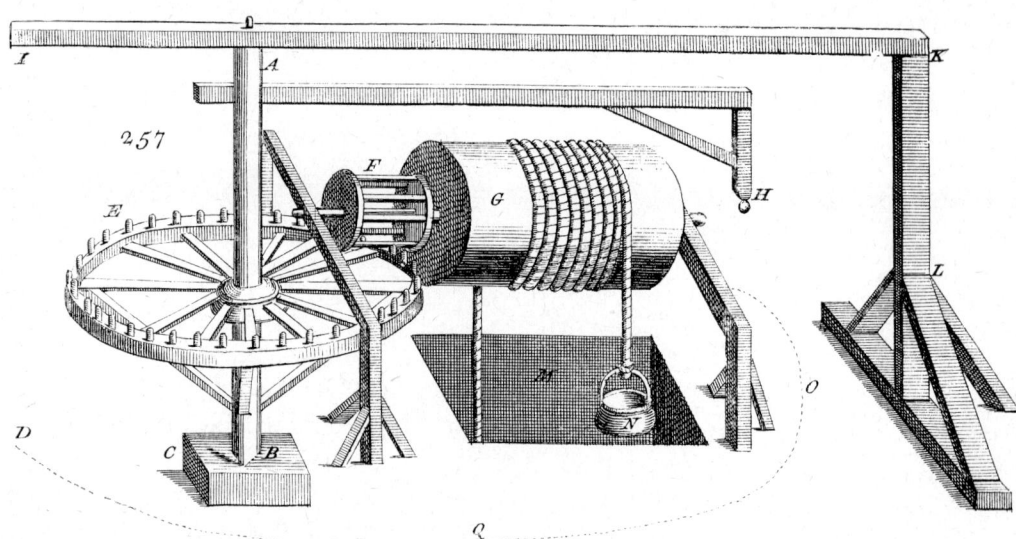

Fig. 24. A Cog and Rung Gin, 1758.[2]
(From Emerson's *Principles of Mechanics*.)

---

[1] Atkinson, F., 1960: 38.   [2] See also Atkinson, F., 1965.

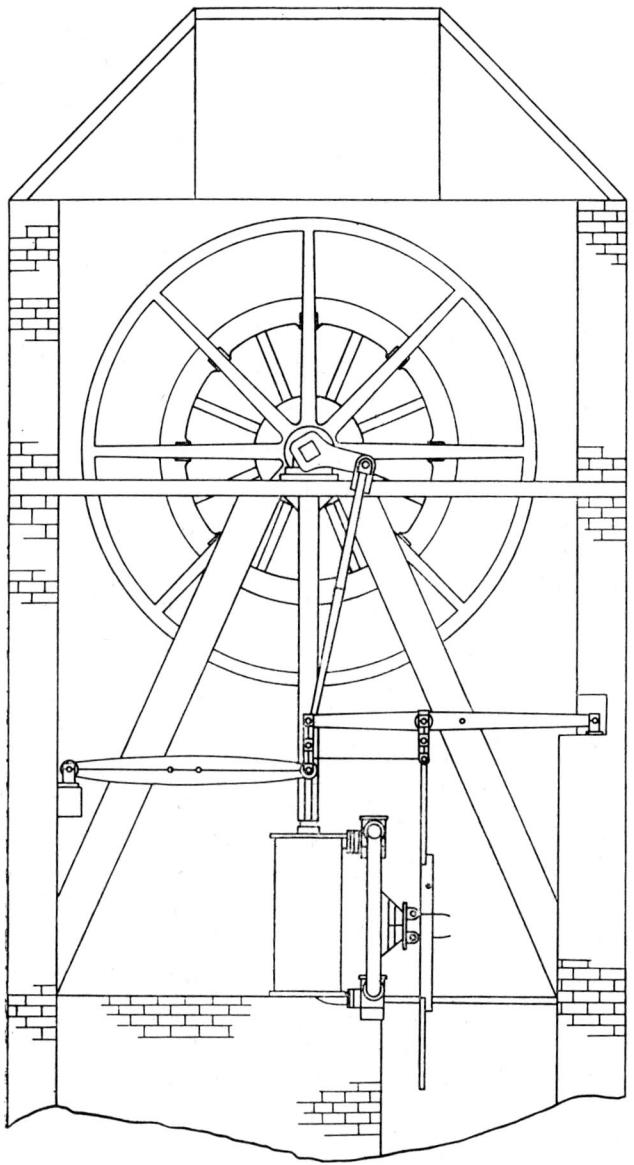

Fig. 25. Elevation of the 'Isabella' Winding Engine.
Above the vertical cylinder are beams to guide the piston vertically, above these the winding drum. (From Atkinson, P., 1956.)

Fig. 26. A Vertical Winding Engine, Built in 1855, at Beamish Colliery.

(cf. Fig. 25). Ceased work, 1962. Note wooden headstock above shaft. Reserved for the Northern Regional Open Air Museum.

Fig. 27. Watercolour of a Cog-and-Rung Gin, c. 1775.
By John Bewick, younger brother of Thomas Bewick, wood-engraver.

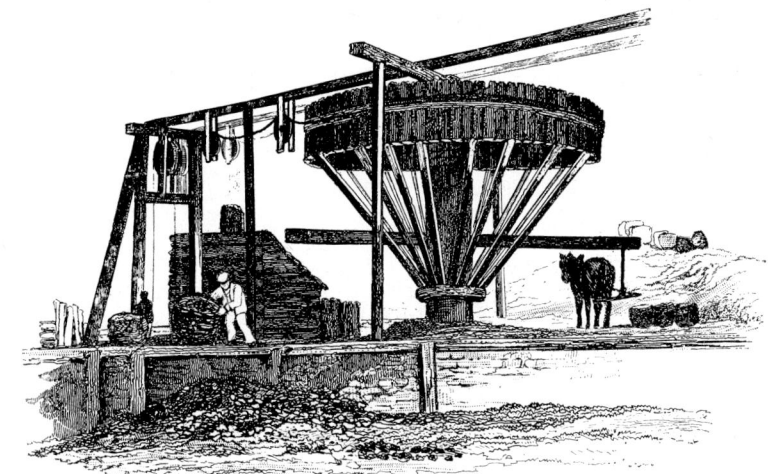

Fig. 28. A WHIM GIN, 1844.

The whim gin succeeded the cog-and-rung gin (Fig. 23), and the horse-track was no longer *round* the shaft. (From Hair's *Coal Mines*.)

Although the horse whim gin continued in use at smaller collieries almost to the end of the nineteenth century, ingenuity was brought to bear on the problem of harnessing the steam engine and during the 1760's and 1770's several curious methods were tried. In about 1780 a Mr Thomas Hunt used the steam engine to pump up water which then drove a water wheel, and this in turn wound the rope. A note in the *Universal Magazine* for February, 1782, illustrates the machine, and informs us that two were erected: at Killingworth Moor Colliery and at Walker. This was one of many such experiments and before the end of the eighteenth century the water wheel became the common method of winding at the larger collieries. In 1797 John Curr estimated that thirty or forty of these water coal gins with their fire-engines were at work in the north, but within a very few years they too were superseded by the steam winding engine.

The improvements to the steam engine had led by 1782 to the double-acting rotative engine (see *Drainage*) and this was soon applied to colliery winding. In 1800 Phineas Crowther, of Newcastle-upon-Tyne, patented a design for a vertical single-cylinder engine which dispensed with the usual beam.[1] Since Crowther was a Newcastle man it is probable that this type of engine was used in Durham soon after its inception. Certainly it eventually became a standard type in the area and remained so long after double-cylinder engines had become accepted practice elsewhere. Fig. 25 shows the mechanism of a typical engine: the Isabella engine at Elemore Colliery, Hetton-le-Hole, Co. Durham, which was probably constructed about 1826. It will be seen that the winding drum was placed directly above the cylinder, thus accounting for the characteristic tall shape of these engine-houses.

The firm of Thomas Murray, of Chester-le-Street, built many Crowther-type engines, as also did J. & G. Joicey, of Newcastle-upon-Tyne. It is their No. 20 of 1855 which is still preserved at Beamish No. 2 Pit, where it was in use until 1962 (Fig. 26). Outwardly the most distinctive feature of the vertical winding engine was its engine house: square and tall, with the ropes running almost horizontally to the pit-head pulleys. Several engine-houses can still be found, though all are now stripped of their original machinery except for that at Beamish.

[1] Watkins, 1955: 205.

Fig. 29. A Whim Gin, 1765.
(From the *London Magazine*, January, 1765.)

Surface Treatment

At the collieries of the eighteenth century a *banksman* stood at the mouth of the shaft and simply unhooked the corves which were then dragged away and tipped. We have a delightful engraving (Fig. 29) from Chester-le-Street, published in 1765, showing the operation as it was still carried on. Across the shaft was fixed a beam (the *striking deal*) on which the banksman could lean whilst *striking the corf*, or swinging it away from the shaft. In his hand he held a long hook with which to reach the rope and pull the corf to the bank. The corf was then dragged by horse-drawn sledge to the coal-heap, where it was sorted. Here are shown four individuals employed to take refuse out and carry it to the adjoining refuse heaps. Two *fire lamps* are shown, burning coal to give a smoky light "for letting the gin-horses, drivers, and banksman see to do their work, as the pits usually work in the night-time".

About 1760[1] the coal owners of the Great Northern Coalfield adopted the practice of *screening* or sieving the coals. This was first introduced at collieries producing coal of inferior quality, for though it involved the waste of much small coal, it enabled the remainder to meet that of the richer mines on terms of equality in the market.

Doubtless these first screens would be of wood, but later metal bars or grates came into use. Moreover, in order to save labour, it became the practice to raise the corves, whilst still attached to the winding rope, to a point well above ground level. From such a raised platform, the corves could be tipped over the screens; the coals coming to rest eventually in a horse-drawn wagon below. This elevated framework of wood or iron at *bank* (*i.e.* above the shaft mouth) was, and still is, known as the *heapstead*: a reminder of the time when the coal was heaped close at hand. Fig. 7 shows this clearly at Hebburn Colliery, and the screens can also be seen in the structure at the right.

An early nineteenth-century writer tells us: "Connected with every pit in the neighbourhood of Newcastle is a contrivance for screening the coals. In most cases it consists of a platform sloping at an angle of about 45 deg. from the raised bank about the pit toward the ground. At intervals are inserted grates 12 or 15 feet in length and about four feet wide, having the spaces between the bars more or less considerable according to the size of coal required to pass through. On each side of these grates, boards confine the coals in their course, and they are likewise boarded underneath, in such a manner as form the surface of spouts, by means of which the dust and small coal which pass through the bars fall into wagons placed for that purpose, as other wagons are placed outside of them for the reception of the screened coal".[2]

When there was (in the eighteenth century) still a use for cheap small coal, for example for saltmaking, wagons would certainly be placed beneath the screens for its reception. But as saltmaking ceased the small coals were themselves further screened. Greenwell[3] states that the small coals, screened out from the *round coals*, were drawn up an incline in an *apparatus tub* to a second screen. This tub was drawn up by a chain or rope connected to the winding engine and was "ingeniously contrived to discharge itself when

[1] Ashton and Sykes, 1929: 194. 213. [2] Holland, 1835: 209.
[3] Greenwell, 1888: *vide Apparatus.*

at the proper height on the apparatus". This second screening separated the small coal into *nuts* and *duff*, and the latter was taken to the waste heap.

Matthias Dunn writing in 1848 illustrated "the plan of a modern coal pit in the County of Durham" (Fig. 30). In the lower part of the illustration are shown the *skreen* arrangements, *hh* were bars of iron $\frac{1}{2}$ in. apart, *kk* were hoppers for the reception of small coal which had passed through the skreen, having outlet at *l*. The large coals before passing into the chaldron wagons *xx* were cleared from bad coal and stones at the platform *mm*. The whole apparatus was enclosed and covered by a roof of wood or slates *nn*. A general view of this kind of structure can be seen in Fig. 7.

Another illustration (Fig. 32) shows nineteenth-century screens in action with the tipping mechanism, or *kickup*, at the top. This was an iron cradle placed at the head of each screen and horizontally pivoted. It was so balanced that when a full tub of coals was run into it, the cradle and tub turned over to discharge the coal. After this the cradle was righted and the tub run off and taken back to the shaft. Also shown in the illustration is the *trap*, just below the kickup. This was a door worked by a handle by the screenman, by which "the coals from the last tub teemed on the skreen are prevented from passing down until the previous tub has been disposed of, and by which door, in proportion as it is raised, the skreenman can control the rush of the coals down the skreen".[1]

As quantities of mined coal increased, through the second half of the nineteenth century, some speeding-up of the screening process became essential. Care was necessary, for increasing the slope of the screens might result in further breakage and a greater proportion of small coal, whereas to increase the number of screens meant an increase in labour costs. For such reasons screens worked mechanically came into operation. Eccentric action produced sufficient movement to speed the separation without damaging the coal and such *jigging screens* were quickly adopted.

After screening, the coals still had to be examined in order that unwanted stone could be picked out, and the large quantities being dealt with in the second half of the nineteenth century necessitated great numbers of boys being employed as coal-pickers. Fig. 31 shows metal sectional *belts* in operation in 1901 at St. Hilda Colliery, South Shields. The belts come from the screens and as the coal passes in front of them, the boys standing at each side pick out stone, shale, and 'brass' (iron pyrites) and throw this into the centre section of the belt. From here this waste would eventually be tipped separately from the coal. This manual coal-picking began to be replaced in the late nineteenth century by coal-washing, partly to speed and cheapen the process, partly because below a certain size it was impracticable to pick out the dirt mixed with the coal. The principle applied to coal-washing had already been used for many centuries in metal-ore dressing, namely that bodies of differing specific gravity fall through water at different velocities, for a given size. Therefore by the use of flowing water, particles of coal and shale could be separated.

[1] Greenwell, 1888: *vide Kickup* and *Trap*.

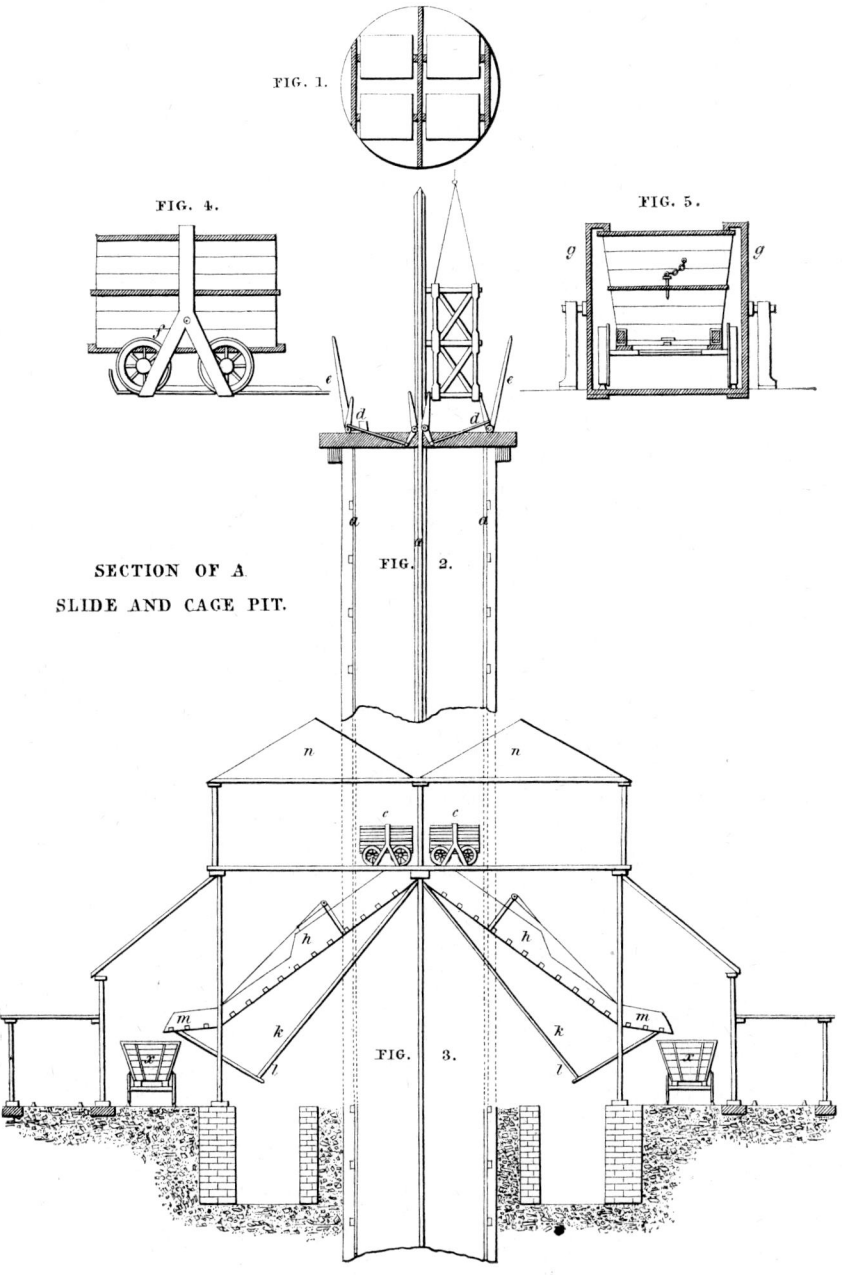

Fig. 30. Pit-head Arrangements, including Screens, 1848.
(From Dunn's *Winning and Working Collieries*.)

Accidents

The life of the miner was, and still is, a hazardous one. Rock falls at the working face, water-flooding, explosion and fire, suffocating gases, and blockage of the shaft have all occurred many times. However, it was as a direct result of certain accidents that working conditions were improved, and two notable examples will be quoted.

'Firedamp' or explosive gas has always been one of the greatest hazards, particularly in mines around the Tyne, and in early writings we frequently find the term 'fiery pits'. During the eighteenth and nineteenth centuries explosions occurred frequently and many caused large numbers of fatalities. A dozen of the more notable are: 1708 Fatfield (69 deaths); 1767 Fatfield (39); 1812 Felling (92); 1815 Newbottle, Fence Houses (51); 1821 Wallsend, North Shields (52); 1835 Wallsend (102); 1844 Haswell, Newcastle (95); 1860 Burradon, Durham (76); 1866 Pelton, Chester-le-Street (24); 1880 Seaham (164); 1882 Trimdon Grange (74); 1886 Elemore, Durham (28).

The Felling Colliery disaster of 1812 so impressed the Rev. John Hodgson of that parish that he widely publicised the accident.[1] In 1813, as a direct result, the "Society for Preventing Accidents in Coal Mines" was formed at Sunderland. John Buddle at once wrote giving a detailed account of the various systems employed in the ventilation of collieries, and this was published as the first Report.[2] The Society then approached Sir Humphrey Davy, who expressed great interest and visited the area. By October, 1815, he had constructed a safe lamp. Yet the safety lamp did not immediately have its desired effect, partly on account of the more fiery parts of the pits now being accessible with supposed safety, and partly due to foolhardiness on the part of the pitmen. John Buddle in 1829 informed a Lords' Enquiry: "They have been fined, and the magistrates have sent them to the House of Correction for a month, yet they will screw off the top of the Davy and expose the naked flame".

A more unexpected tragedy took place on 16th January, 1862, at Hartley Colliery, when 204 men and boys died in one of the most appalling mining catastrophes in the annals of this country. The colliery at New Hartley, in south-east Northumberland, was worked with only one shaft in which was also fixed a set of pumps, the pump 'shaft' being separated from the winding 'shaft' by wooden brattices which extended from top to bottom. The beam of the pumping engine, the largest and most powerful in the north and weighing more than forty tons, suddenly and without warning broke. About half of it plunged down the shaft, tearing away the brattices, ripping off the stone and timber that protected the walls, carrying pipes, gearing, and hundreds of tons of debris in its descent. The accident happened just after the back-shift men had gone down to relieve the fore-shift. Six days elapsed before a way could be made through the debris and by that time all had died, suffocated by poisonous gases and lack of fresh air.

A large public meeting was held in Newcastle nine days after the accident and a petition drawn up demanding that two shafts to every pit be made compulsory by law. As a direct result of such agitation an Act was passed to this effect later that year.

[1] Hodgson, 1813.   [2] Buddle, 1814.

Fig. 33. A Last Message Scratched with a Nail on a Tin Water-bottle. Seaham Explosion, 1880.

The message reads:

Dear Margaret,

There was 40 of us altogether at 7 a.m. Some was singing hymns, but my thoughts was on my little Michael that him and I would meet in heaven at the same time. Oh Dear wife, God save you and the children, and *pray for me.* . . . *Dear wife Farewell. My last thoughts are about you and the children. Be sure and learn the children to pray for me. Oh what an awful position we are in.*

Michael Smith, 54 Henry Street.

(From *Explosions in Coal Mines*, W. N. and J. B. Atkinson, 1886.)

THE PITMEN

Housing

In common with so many other everyday conditions, little has been recorded about life in a pit village in the eighteenth century. A few photographs have been taken of cottages of the period, but of course in occupation under nineteenth-century or even twentieth-century conditions. Contemporary artists rarely troubled to record such menial matters and a rare example is the little drawing by Grimm (1734-94) of pitmen's "houses on the road to Newcastle" (Fig. 35).

For the early nineteenth century we are obliged to study evidence published in Commissioners' Reports. Thus from 1840 we read: "Within the last ten years collieries have been opened in very many places between the Wear and the Tees; and wherever a colliery has been opened a large village or town has been instantly built close to it, with a population almost exclusively of the colliery people, beershop people, and small shopkeepers. The houses have either been built by the colliery proprietors, or have been so by others, and let on lease to them, that they might locate their people". The village of Coxhoe was quoted as typical, extending about a mile along both sides of the public road, with a break every ten or so houses to make a thoroughfare to the streets running off left and right. These cottages were built of stone plastered with lime, with blue slate roofs, "as like to one another as so many soldiers". They had no yard in front, nor behind, and comprised two rooms and a semi-attic bedroom. The ground floor was made of clay, sand, and lime. The whole expense of erecting such a cottage was £52 and it was rented for £5 a year. In some areas the houses had nearby pigsties, and there are references to little brick buildings in the streets, designed as public ovens.

The population of this village was estimated at 5,000, the workpeople of several collieries living there. Altogether there were thirty beer-shops in the village, and although at that time there was no Church of England church or chapel, the Wesleyans and Primitive Methodists had established their meetings and had many adherents.

One witness in 1840 commented on a remarkable dissimilarity between the cottages and their furniture, which he thought to be peculiar to the homes of the northern pitmen: he found the houses to contain comparatively showy and costly furniture. Typical items of furniture in the principal room were an eight-day clock, a good chest of drawers with brass handles and ornaments, reaching from floor to ceiling, and a fine four-post bedstead with a large coverlet composed of squares of printed calico. These last two were often of mahogany, and were deemed indispensable by a decent newly-married couple. They were paid for by instalments. Also mentioned in these descriptions are the *dess-beds*, (sometimes called desk-beds): one would often be in the principal room for the youngest of the family. The rest of the family would be distributed in the back room and attic.

The dess-bed and a somewhat similar bed, the *chiffonier*, were necessary items in a crowded household. They folded away inconspicuously during the day and were quickly available for the night. Such beds remained in common use well into this century.

Fig. 34. Cottages at Ryhope, Co. Durham, in 1961.
Note double doors once typical of such houses.

Fig. 35. Pitmen's Houses near Newcastle, 1778.
Sketch by S. H. Grimm (British Museum).

Fig. 36. A Pitman's Cottage in 1963 at Shiremoor, Northumberland.
Left of the fireplace is the *side-boiler* for water-heating; right is the circular oven. A brass strip is nailed along the edge of the mantelpiece and beneath is a brass rail for airing clothes. See also American alarm clock, glass candlesticks, and Sunderland-ware plaques. There was no sink.

45

Fig. 37. A West Durham Pitman, about 1900, Carrying a 'Midgy' Lamp.

Jacket photograph is of a group of pitmen near Crook, Co. Durham, about 1890: eldest (second from left, middle row) wears a soft cap somewhat earlier than 1890. This, and Fig. 37 are in the collections of the Northern Regional Open Air Museum.

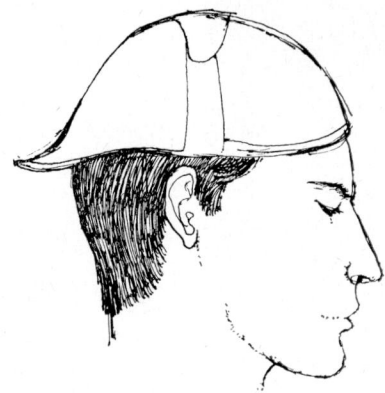

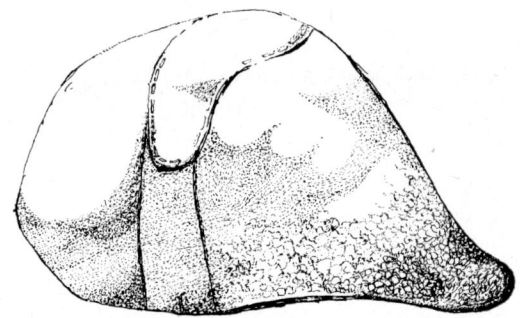

Fig. 38. A Nineteenth Century Leather Helmet Worn by Deputies.

From Leasingthorne. Now in the collections of the Northern Regional Open Air Museum.

Clothing

The Durham pitman was noted in the nineteenth century and early twentieth century for his characteristic clothing: boots and heavy woollen blue stockings and breeches cut off at the knees and split at the ends so that they could easily be drawn over the pit boots. The hewer also wore a woollen waistcoat of natural rough felt and collarless. This he wore on top of his shirt and under his jacket whilst going *in bye*, on account of the cold air intake. At work he would strip to the waist and afterwards wear the waistcoat next to his skin to absorb the sweat.

J. R. Leifchild (1850) wrote of hewers: "Their pit dress is made entirely of coarse flannel: a long jacket with large side-pockets, a waistcoat, a flannel shirt, a pair of short drawers, and a pair of stout trousers worn over them. Add to these a pair of *hoggers* or footless worsted stockings, a tight-fitting leather cap, and you have the hewer ready for the pit. A pit suit costs about one pound, though some wives and daughters can make them". Cloth caps would be worn, but never a helmet until well into the present century. Recollections of elderly miners to-day are that leather caps or helmets were worn by deputies or overmen, but never by the pitmen.

This is also the place to comment that the word *pitman*, now rarely heard, is recollected as having been the general term in this coalfield until a few years ago. The official term, miner, has now replaced it.

Wages

Some idea of wage structures and working conditions is given by a complete "Account of work" for a fortnight in 1802, reprinted on p. 69, from East Kenton Colliery, Northumberland. Twenty-three hewers are listed and the amount of coal hewn each day is shown in corves. This is totalled for the period under the heading *Round Coalls* in scores and corves. (Thus Jos. Hunter hewed 148 corves, or 7 scores and 8 corves of 'Round Coal'.) In addition a certain amount of 'small coal' was *stowed underground* and this was paid at the same rate of two shillings per score. Other special payments were for *wet working*, for working *Rambel*, for working *Doulle Boredes* (double bords), and for working under the *Tope coal*. Ramble is a thin stratum of shale often found immediately above the coal. Greenwell[1] states that this falls and is troublesome to separate from the coal. The extra payment for working double bords was to compensate for the inconvenience of two men working in one bord or place of working.

Twelve putters were engaged, together with four horse-drivers and three trapper boys. The overman also itemised his totals for various tasks such as *shovling tram Road* (clearing the underground route), *attending Machine at tope of 20 Fathom Pit* (probably a horse-gin for winding coal out of the pit), and *Srowing Propes and Plancks* (sawing props and planks for supporting the roof of the workings). The *Chearity Muney* was probably a small fund maintained for payment in case of accidents or death.

[1] Greenwell 1888: *vide Ramble*.

SURFACE TRANSPORT

Horse Wagonways

We have seen how the proximity of easily worked coal outcrops to the riverside was the major cause of early coal workings along the Tyne. So it was a natural development that as these workings became exhausted, coal should be brought by horse and cart from collieries slightly further from the river. Such distances grew and means of speeding and so cheapening the journey had to be found. In the mid-seventeenth century, wooden wagonways were laid over which wheeled vehicles could more easily be drawn. These were a great improvement on the muddy, often almost impassable cart ways, and a horse could pull a large four-wheeled wagon, usually holding about two tons. Two sets of wooden rails were laid, and in descending one track the driver hitched his horse to the rear of the wagon, unless the descent was too gradual for the wagon to travel by its own weight: in ascending the other track the animal pulled back the empty vehicle (Fig. 44).

Various means were adopted for keeping the vehicle's wheels on the wooden planks or rails (which were of oak, ash, or birch), but by the 1670's flanged wheels were in use. In 1676 Roger North saw "rails of timber, from the colliery down to the river, exactly straight and parallel, and bulky carts are made with four rowlets fitting these rails; usually the carriage is so easy that one horse will draw down four or five chaldrons of coal".

Owners of land over which the wagonways passed naturally demanded recompense, and the financial problem of 'wayleaves' was a vexed one which recurs frequently in business correspondence. Roger North, in the description quoted above, mentioned that the owner of a quarter of an acre of land expected to receive £20 per annum for wayleave.

The end of the seventeenth and beginning of the eighteenth century saw a great increase of coal working and wagonways. After the Great Fire of 1666 and the rebuilding of London's wooden houses by brick houses with chimneys the demand for house coal increased and fortunately the coal around the Tyne was perfectly suitable for this purpose. By 1700 coal was frequently carried eight or ten miles to the river from collieries at Tanfield, Pontop, and South Moor, and it was not until the nineteenth century when the steam railway again reduced the costs of carriage that collieries were sunk much further inland.

An inspiring description of this area published in 1776 (but probably written several decades earlier) says: "We saw Col. Lyddal's coal-works at Tanfield where he carries the road over valleys filled with earth, 100 feet high, 300 foot broad at the bottom; other valleys as large have a stone bridge across, and in other places hills are cut through for half a mile together; and in this manner a road is made . . . for five miles to the river side, where coals were delivered at 5s. the chaldron".[1] The bridge was of course the Tanfield Arch, or Causey Arch, as it is known locally, which was built in 1727 for Colonel Liddell, one of the "Grand Allies", by Ralph Wood, a local mason (Fig. 39). Wagonways spread across the coalfield during the eighteenth century and a useful map of 1787 (Fig. 43) shows their routes.

[1] Stukeley, 1776: II, 69.

Fig. 39. CAUSEY ARCH, THE EARLIEST RAILWAY BRIDGE, ERECTED IN 1727. Also called Tanfield Arch. Scheduled as Ancient Monument.

Fig. 40. *Left:* A WAGON SKETCHED IN 1724. (From Sir John Clerk's *Journal*.)

Fig. 41. *Below, left:* A CHALDRON WAGON, SKETCHED ON A MAP OF 1749. (From estate map of Longbenton, Northumberland.)

Fig. 42. *Below, right:* A WOODCUT OF A HORSE-DRAWN WAGON, 1835. Note brakes and brake lever. (From Holland's *Fossil Fuel*.)

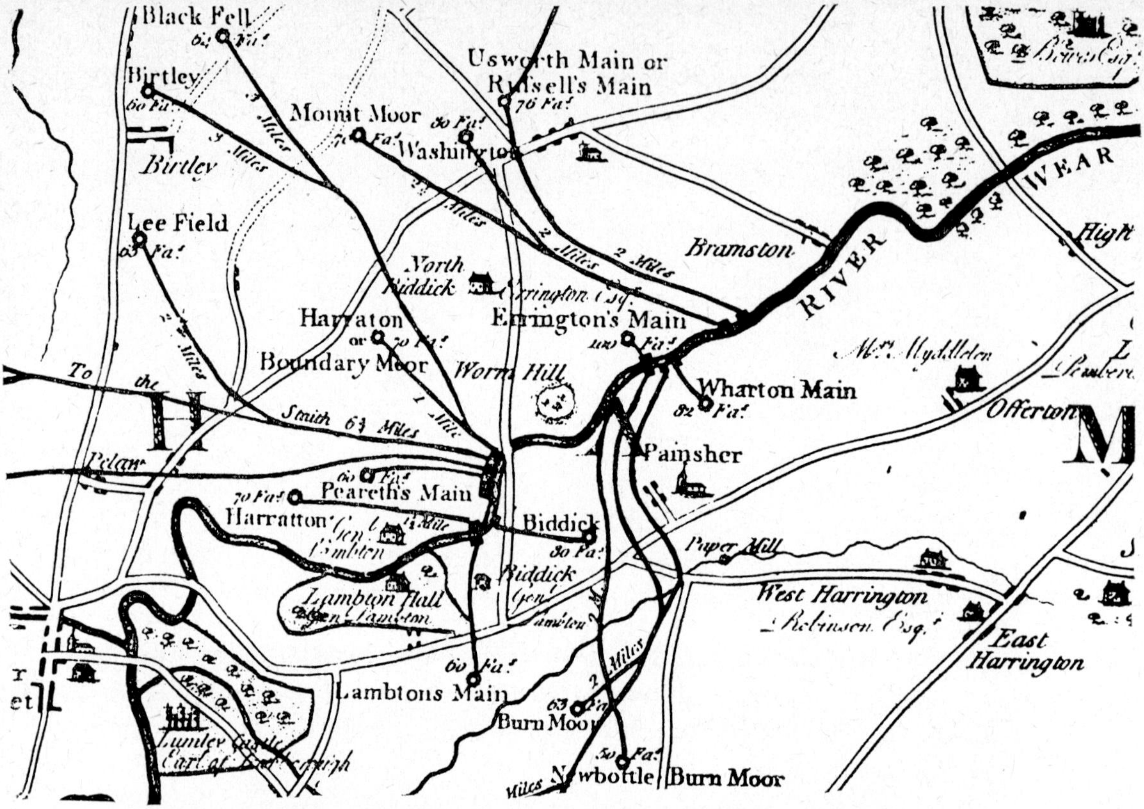

Fig. 43. Part of a Map of 1787 showing Horse-wagonways.
This part is reproduced full size. (From Gibson's *Plan of the Collieries of the Tyne and Wear*.)

Fig. 44. A Chaldron Wagon, 1773.
Covered staithe and keel in background. (From Morand's *L'Art d'exploiter les Mines*.)

The actual tracks of the wagonways continued to be made of wood into the nineteenth century and when John Bailey travelled County Durham to write his *Agricultural Survey* in 1810 he commented that wooden ways were still used extensively on Tyneside and cost 5/- per yard or £440 per mile, but that the substitution of iron rails was occurring. A change parallel to that of complete replacement of wood rails by metal rails, was the covering of existing oak rails by flat cast-iron plates. A remarkable continuance of this practice came to light as recently as 1965, when a firm of timber merchants near Sunderland dock was found still to be using heavy oak rails plated with iron.

To trace the forms of rails, the fish-bellied and the bull-headed, and details of their chairs and sleepers is a specialist task, and one which still remains to be completed. We may merely add that the introduction of the familiar fish-bellied rail, used in later years by George Stephenson, probably took place in the 1790s.[1] So far we have merely mentioned that chaldron wagons ran along these wagonways. It remains to consider the shape, size, and development of these vehicles. One of the earliest drawings we have was jotted down in his Journal by Sir John Clerk, a Scot who visited the Newcastle coalfield in 1724. Travelling south, he came to a colliery or 'coalwork', as he called it, about five miles south of Morpeth and two miles to the east of the high road to Newcastle. From here a 'cassey way' of about $4\frac{1}{2}$ ft wide and four miles long carried the coals to the sea. These coals, he wrote, "are drawn on a kind of wagon with 4 thick wheels of solid timber unshod and by one horse. The wagon is of this form (Fig. 40) holding a quantity of coals near to a Scots chalder".[2]

The next good illustration we have was made some twenty-five years later (Fig. 41). This was of a wagon in use at Long Benton, Tyneside, in 1749. The wheels are still probably of timber, but with four massive spokes, and this time a brake is shown: a curved bar over one of the back wheels. By the mid-eighteenth century the two fore wheels were being made of cast iron, and weighed several hundredweight; whilst the two rear wheels were of wood and had the iron axles fixed in the wheels, turning with the wheels.[3] A print of 1768 shows such a wagon running to the staithes with the horse following (Fig. 44).

From the mid-eighteenth century to the mid-nineteenth century, the shape of these wagons changed little, except that all four wheels were eventually made of cast iron, when engine haulage changed the character of the runs and braking on the rear wooden wheel was no longer necessary. We have some precise details because in 1851 two tiny model wagons were made at Blyth colliery, for showing at the Great Exhibition. Afterwards they passed into the colliery engineer's family and were used by several generations of children until one was finally rescued and is now in the Regional Museum collections. Later examples of actual wagons, probably dating from about 1870, are still in use, in rapidly diminishing numbers, at Seaham Harbour. Their design is somewhat different, since each company had developed its own peculiarities of shape and size, and their capacity is naturally greater than those of the early horse-drawn wagons, but the line of development can clearly be seen.

[1] Lee, 1960: 13.   [2] Atkinson, F., 1965: 427.
[3] Letter to the *London Magazine*, March, 1764, and illustration.

Steam Haulage

The horse-drawn trucks brought coal to the River Tyne and the River Wear along wagonways which, by the eighteenth century, might be several miles in length. Earlier ways of cheapening and speeding their journeys had included laying tracks of gradually improved quality. In 1805 came a complete innovation: the replacement of the horse by the stationary steam engine.

We have already seen how the atmospheric engine was gradually improved through the eighteenth century until by the end of the century it was supplanted by the steam engine for colliery winding. It is therefore not surprising to see this next experiment: if a steam engine could wind coal *up* a shaft, could it not haul trucks of coal *along* an incline? Before long the wagonways of the northern coalfield were being transformed. John Curr, a coal viewer or engineer of Sheffield, in about 1805 was the first to use engine power to raise coals from the valley at Birtley, near Gateshead, to the heights of Black Fell. This was almost immediately followed (1808) by a great scheme of Messrs Harrison, Cook & Co., to convey coals from Orpeth Colliery to the Tyne over the heights of Eighton Banks, by a succession of inclined planes partly worked by fixed steam engines. Unfortunately, although effective its cost was greater than anticipated and the company was ruined.[1]

Some re-alignment of tracks was necessary when horse wagonways were converted, because horseways tended to follow the land along gentle winding routes whereas straight runs and direct inclines better suited the stationary engine.

Frequently the engine was sited tangentially on a bend, in such a way that trucks could be given a direct haul up the incline, then run "over the hump" at the top and started on their downward journey to the river by way of a self-acting incline. This kind of route, when studied on a contoured map, can often identify probable engine-house positions even when there is no indication on the modern map. Indeed the map has yet to be drawn of County Durham, showing such routes and collieries served with haulage-engines and their dates of construction and use (Fig. 45). The self-acting incline was a refinement by which, on a downward run to the river or sea, a set of loaded trucks could haul back by their own weight a set of returning empties, the sets passing at a midway loop on most lines. The wire rope connecting the two sets of trucks ran round a large horizontal pulley placed below track-level and the speed of descent of the full trucks was controlled by the brakesman at the incline head.

One by one, as they have gone out of use, the engines and engine-houses of the north-east have disappeared, although a few empty semi-derelict engine houses can still be found. At Warden Law (on the Hetton to Sunderland line) was the last stationary steam haulage engine to operate on a colliery line in County Durham. This line was constructed in 1822 by George Stephenson, though his engine at Warden Law was replaced by a somewhat larger one in 1836, made by Murrays of Chester-le-Street. In 1959 this line was finally closed and the engine has been dismantled and is now stored for the Regional Open Air Museum.

[1] Dunn, 1844: 52.

Fig. 45. THE STANLEY HAULAGE ENGINE, NEAR CROOK, CO. DURHAM, BUILT IN 1857.
Boiler house on right. At right of engine house is small lean-to, to house winding drums.

Fig. 46. LOOKING DOWN PELTON SELF-ACTING INCLINE.
In the distance there are three rails which divide towards the foreground, and the tracks come over 'the humps'.

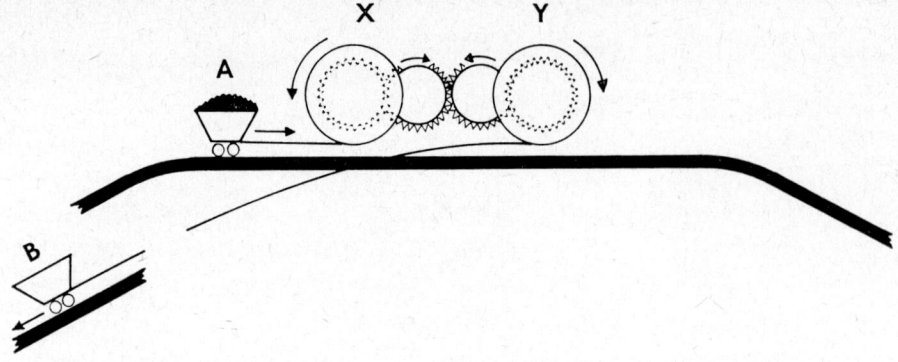

Fig. 47. THE WINDING SEQUENCE: WARDEN LAW.

1. A loaded set of wagons A is hauled to the crest of the slope by drum X, whilst an empty set B descends, controlled by drum Y. (The drums are interlocked by large gearwheels.)

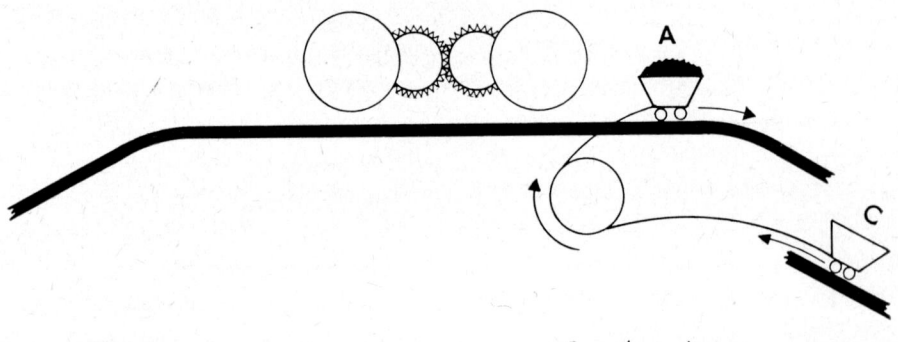

2. The loaded set A is now disconnected from the haulage cable and attached to the cable of the gravity-controlled system. A is now allowed to descend the incline to the docks, thus drawing up empty set C to the crest.

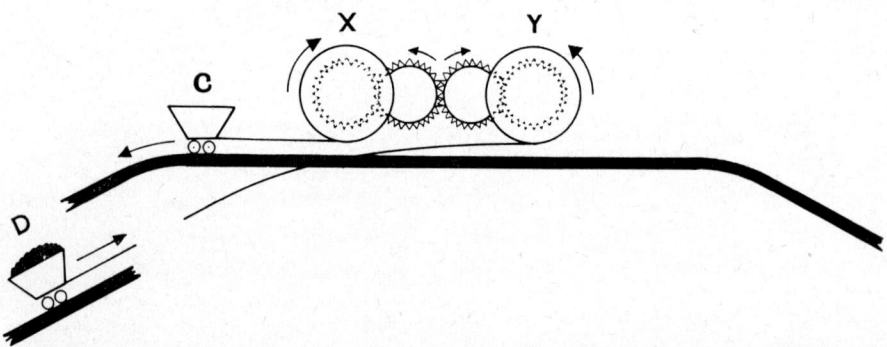

3. Set C is then connected to the cable-link from which A was disconnected. By this time the empty set B will have been exchanged for a loaded set D. The engine is now put into reverse motion so that loaded set D is hauled to the crest, from the colliery, whilst lowering empty set C.

Railways

Most of the haulage inclines of the early nineteenth century were eventually replaced by locomotive-powered railways, first pre-occupied with the passage of coal and lime, and later sometimes carrying passenger traffic. Four principal lines utilised all available means of haulage, including early locomotives, and were eventually operated in some cases by electric winders and diesel locomotives.

The Stockton and Darlington Railway was promoted to provide an outlet for coal from south-west Durham. Originally a canal had been proposed in the hope that the Tees might compete with the Tyne and Wear, but this scheme was replaced by a plan for a tramroad, and this in turn by the proposal put forward by George Stephenson. The route of this new railway ran from Witton Park Colliery by Etherley Incline to St Helen's Auckland and thence by Brusselton Incline to Shildon. At the top of each of these inclines stood a stationary steam winding engine and remains of Brusselton engine house may still be seen converted into a cottage. Near West Auckland, Stephenson spanned the River Gaunless with the first cast iron railway bridge, a portion of which is preserved in York Railway Museum. At Shildon steam locomotives took over and brought the coal to Stockton Quay. The first locomotive to run on a public railway, *Locomotion No. 1*, is preserved at Bank Top station at Darlington (Fig. 48).

The Stanhope and Tyne railroad, opened in 1834 (although parts of the route had been in use since 1670), was designed to carry coal and lime from north-west Durham to South Shields and negotiated a most difficult terrain west of Birtley. At the Stanhope terminus were Ashes Quarry, lime kilns, and a coal depot. From here wagons were drawn up two inclines by stationary engines at Crawleyside and Weatherhill. Horses, Meeting Slacks stationary engine, Nanny Mayor's self-acting incline, and then further horse power brought the traffic to Hownes Gill, one and a half miles south of Consett. This dry ravine 800 ft wide and 160 ft deep was originally overcome by two lift planes having a common stationary engine at the bottom. On these steep lift planes were cradles having wheels of unequal diameter in order to carry the wagons at the horizontal. In 1858 this curious installation was replaced by an elegant viaduct, still in use, built of Pease's Roddymoor brick. Traces of the lift planes may yet be seen in the ravine, on the north side of the viaduct.

From Hownes Gill engines, self-acting inclines, horses, and locomotives worked the traffic to three large coal drops at South Shields. The entire route from Stanhope to South Shields was worked by: horses ($10\frac{1}{2}$ miles), stationary engines (9 : 11 miles), self-acting inclines (5 : 3 miles), and locomotives ($9\frac{1}{4}$ miles). Several buildings remain along the route of this particularly interesting line, one of the best preserved being the coal and lime depot dated 1834 at West Bolden level crossing (Fig. 49).

The Tanfield wagonway was constructed in 1727 (see *Horse Wagonways* above). The line became a railway in 1839, when its conversion cost the Brandling Junction Railway £37,000, and it ran via three main inclines to 'drops' at Redheugh. Finally, it ended at Dunston staithes, to which the last portion closed on 18th May, 1964.

Fig. 48. GEORGE STEPHENSON'S 'LOCOMOTION', 1825.
Preserved at Darlington (Bank Top), B. R. Station.

Fig. 49. LIME AND COAL DEPOT AT WEST BOLDON, BUILT IN 1834.
Stanhope and Tyne Railway.
Reserved for the Northern Regional Open Air Museum.

Staithes

The wharves by means of which the coal was shipped from wagon to vessel were locally known as *staithes*. The term is still used, though the purpose and form has changed. The banks of the Tyne, later the Wear, and eventually such ports as Seaham Harbour, were the shipping points to which the wagonways led. These staithes originally served not merely as gangways for loading, but also as storing places for coal. The mine owners found it desirable to place a large part of their stock on the staithes, ready to ship at once, rather than on the 'heaps' beside their pits, because shipmasters did not relish delaying their sailings while small cart loads were brought from the colliery. Considerable stocks were so maintained, and records for Newcastle in 1640 suggest that as much as 200,000 tons might be waiting for sale at the six or eight principal staithes on the Tyne. Such staithes were covered with a roof of timber to keep the coal dry (Fig. 43), for water added to its weight.

The original gangways for loading coal were nothing but narrow banks of earth and rubbish extending a few yards from the shore, and labourers with wheelbarrows collected the coal from the stacks and emptied it into small river-boats or *keels* which drew alongside. During the seventeenth century ships entering the Tyne dropped anchor further and further downstream for not only did the size of ships increase, but the river bed became choked with ballast and loose coals from the staithes. The keels were therefore essential. Before the end of the seventeenth century the colliery owners began to build gangways on wooden piles and by the middle of the eighteenth century all the staithes on the Tyne were of this construction.[1]

Daniel Defoe in 1726 wrote "The coals . . . are loaded into a Waggon, which by means of . . . a Waggon-way . . . carries them sometimes three or four miles to the nearest River . . ., and there they are either thrown into or from a great store-house, called a Stethe, made so artificially, with one part close to or hanging over the Water, that the lighters or keels can come close to or under it, and the coals at once shot out of the Waggon into the said Lighters, which carry them to the Ships".[2] As wagonways became more efficient the necessity to store coal at the staithes decreased and means were found during the eighteenth century of delivering coal quickly to the boats. As early as 1730, there were *spouts* at the larger staithes: wooden chutes down which the coals were shot into the keels below.[3] The next improvement appeared around 1800, when *drops* came into use. By these machines the wagons of coal were lowered to the ship's deck before being emptied, thereby causing less breakage of the coal.

By the first quarter of the nineteenth century staithes had reached large proportions. A description of 1826 of the Wallsend staithes (said to have been constructed "within the last 20 or 30 years") states: "a staith is a platform or railway, supported by pillars, jutting into the river in one case 169 feet, and in another 153 feet; the height being 36 feet and the breadth 38 feet. To the end of each staith is attached a moveable cradle and drop which shoot out beyond the fixed staith 45 feet".[4]

[1] Nef, 1932: I, 387.   [2] Defoe, 1726: II. pt. ii, 30.   [3] Hughes, 1952: 256.
[4] Keelman's Trial at Carlisle (British Museum: 1875. d. 13).

Fig. 50. Harton Staithes, South Shields, in about 1880.
Nearest coal drop has wagon about to be lowered.

Fig. 51. Amble Staithes, Northumberland, in 1965.
Chute (seen up) can be lowered to collier.

Fig. 52. COAL DROP AT SEAHAM HARBOUR, CONSTRUCTED ABOUT 1850.
Only known remaining example. Now dismantled and stored for the Northern Regional Open Air Museum.

Various methods of controlling the drops were adopted. Some were counterbalanced by weights over pulleys,[1] others by heavy chains falling into pits. Scores of these machines operated along the Tyne and Wear during the nineteenth century (Fig. 50) but all have now gone. Modern staithes can still be seen at Amble and Blyth (Fig. 51) and a few modern shoots still operate on the Tyne, but the only remaining coal drop of the northeast stood until recently at Seaham Harbour, where it was last used about 1900 (Fig. 52). It has been dismantled and is now stored for the Regional Open Air Museum.

Fig. 53. THE SEAHAM HARBOUR COAL DROP.

---

[1] Tomlinson, W. W., 1914.

Shipping

The higher reaches of the Tyne and Wear were too shallow for sea-going vessels and *keels* or lighters were employed for several centuries. They are first heard of in connection with the coal trade in the fourteenth century and went out of use in the nineteenth century. Throughout this period they seem to have changed little in size or appearance. Tyneside keels were carvel-built (*i.e.* with their planks fitted flush), double-ended craft, built almost without sheer (*i.e.* very little upward slope towards bow and stern). They were somewhat oval in shape and six or eight paces in length, with rounded bottoms and almost no freeboard when carrying their normal burden of 21 tons of coal (8 Newcastle chaldrons). William Stukeley describes the operation of keels in his *Itinerary* (1725): "The manner of rowing their great barges (at Newcastle) is also very particular; four men manage the whole, three to a great and long oar, that push it forward, and one to another such a-stern, that assists the others' motion, but at the same time steers the keel, and corrects the bias the others give it". A single mast with square sail was sometimes used, and the keels normally made use of the tide for a loaded journey (Fig. 60). When they had made fast to the *hoy* or *collier*, the keelmen fell to with their bare hands and heaved coal over the sides into the hold, using a shovel only for the small pieces and dust.

In the early nineteenth century, as spouts and coal drops came into common use on large staithes built well into the river, the keelmen complained bitterly.[1] Indeed after about 1780, the collieries that were opened below the Tyne bridges ceased to make use of keels and loaded coal direct into the vessels by means of spouts. Various agreements provided temporary amelioration, and after a 'long stop' or strike in 1822, the unemployed were absorbed by putting an additional man into each keel, and the number of boys employed was reduced. Yet the keel gave way only slowly to the spout, and at no time could spouts be used above the bridges.

The hoy or collier, a seagoing vessel, was built to hold a maximum quantity of coal and to be navigated by the minimum number of seamen. By the time of the Restoration nearly all the old English merchantmen had been superseded on the Newcastle-London run by vessels specially designed for coal transport. The Newcastle collier was probably 60 ft long and 20 ft wide with three tall masts. Except at the prow, where there was a high forecastle and at the stern where there was cabin space for a few sailors, the side came so near to the water that when a keel drew abreast the keelmen had no difficulty in putting coal into the hold through a port-hole. The vessel was manned by common seamen who did not handle cargo, and the keelmen did all the loading.

Under normal conditions the voyage from Newcastle to London took approximately a fortnight, but delays were occasioned in loading and unloading, and taking ballast. At the port of destination a gang of coal heavers came aboard with shovels and baskets. The coal was filled into the baskets which were passed up from the hold and turned into the measure—a Winchester bushel (36 equalling one London chaldron)—in the presence of a 'Sworn Meter'. The coal was then emptied over the side into a waiting lighter.

[1] Ashton and Sykes, 1929: 198.

LOCAL USES

Coking and By-Products

Since about the thirteenth century, when coal was first mined extensively in England, it was known that the residue left on a smith's hearth after heating lump coal had special properties. This residue was not only free from offensive smoke, but it reacted less with hot iron and therefore permitted the smith to work on wrought iron without hardening it. Charcoal, on the other hand, reacted with hot iron and carburised it, producing a steel which was harder to work. The smith only produced small quantities of *coaks*, as they were often described, but these could be made on a larger scale by heaping together large pieces of coal, covering with hay or straw and then loam tamped down with a shovel. In fact this simulated the process of charcoal making. The heap was lit at the bottom and air allowed to draw in around the foot of the heap. When the fire died out the coke was allowed to cool and raked out. This particular method could not be very successfully used with certain Durham coals which swelled on heating and blocked the spaces in the heap.

Nevertheless we find references to *cinders* in early eighteenth century documents and it is clear that attempts were made to produce cokes by burning heaps in the north. For instance, at pits worked on a *tentale* rent (a rent payable for every ten chaldrons of coal sold), it was possible to evade part of the rent by burning *cinders* at the pithead or "in the back lane", for these coals were not then included in the rent. Indeed George Claughton found himself in Morpeth gaol by the middle of the century, faced with a large claim for arrears for doing this. In 1746 he had sixty-one "fires for burning cinders" at Low Fell.[1] A "Plan of the River Wear" for 1737 shows three 'cynder' staithes included among the seventeen staithes on the river, showing that coke making for export was important even at that early date.[2] Difficulties must have been experienced in this "cinder-burning" with the swelling coals of the north and there are references for 1752-3 at Gosforth Colliery[3] to "many tryals to burn it to cynders, but to no effect". By the mid-eighteenth century it was found that the swelling type of coal could be coked in an enclosed oven and the earliest record is for 1759 where there is mention at Chance Pit, near Throckley, Northumberland, to 'Sinder ovens'.[4]

When Gabriel Jars visited England in 1765 he saw nine *beehive ovens* in three batteries of 3, each 10 ft in diameter, on the banks of the Tyne (Fig. 54). These were enclosed brickwork ovens which allowed space for swelling, yet kept out the air which would have completed combustion. Jars describes and illustrates these ovens and states that the distillation yielded an ashen-grey and very porous and light fuel named *cinders*, chiefly used to heat the drying kilns for making barley into malt. Although the manufacture of coke in beehive ovens thus appears to have originated near Newcastle, the production of coke for export was small until the middle of the nineteenth century. Matthias Dunn in 1848 wrote: "Of late years a new and important trade has been opened for the small

[1] Hughes, 1952: 256. [2] Mott, 1936: 31. [3] Mott, *op cit*. [4] Mott, *op cit*.

coal, in the formation of coke, for the use of locomotive engines, ironworks, breweries, etc., at home, as well as for general consumption abroad. Indeed so rapid and important has been the increase of this trade, that some collieries have erected an apparatus for crushing their large coal into small, the better to effect the production of coke".[1]

In the older ovens the waste gases issued from the charging hole in the top of the oven, but by 1860 these gases were collected by a flue common to a row of ovens, and used to heat a waste-heat boiler. In some designs about this time, the effluent gases were used to heat under the floor of the ovens, so giving a saving in coking time and an increased yield. Various attempts were made to utilise the waste gases as a source of by-products such as tar and sulphate of ammonia, and the first battery of by-product coke ovens in Great Britain was erected in 1882 by Messrs Pease and Partners at their Pease's West Colliery at Crook, Co. Durham (Fig. 56). This was a battery of twenty-five Simon-Carves ovens each 23 ft long, 6 ft 6 in. high and $19\frac{1}{2}$ in. wide, a charge of $4\frac{1}{2}$ tons being coked in 60 to 72 hours. A small section of these ovens was preserved *in situ* by the National Coal Board but has now had to be dismantled, and is stored for the Regional Open Air Museum. Despite the obvious efficiency of machine-charged vertical ovens, beehive ovens continued in use for many years, though in declining numbers (Fig. 55). The coke produced by beehive ovens was quite different in character from that of vertical ovens and believed to be superior for many purposes. It was not until 1958 that the last beehive coke ovens ceased to operate in Great Britain. These were at Rowlands Gill, County Durham, and a small section is preserved *in situ* by the National Coal Board.

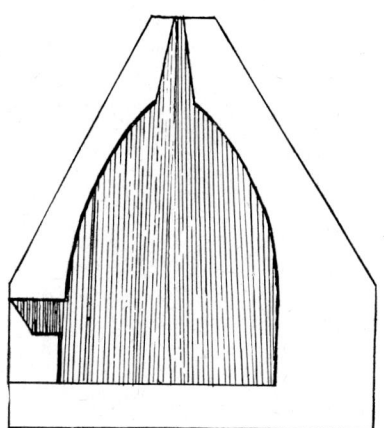 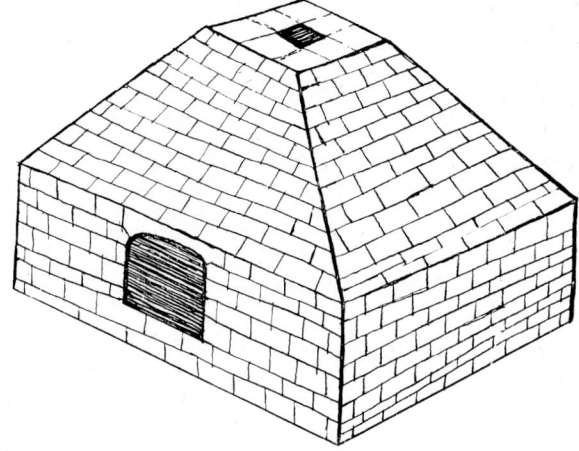

Fig. 54. BEEHIVE COKE OVENS NEAR NEWCASTLE, 1775.
(After Jars: *Voyages Métallurgiques*.)

---

[1] Dunn, 1848: 73.

Fig. 55. BEEHIVE COKE OVENS NEAR ROWLANDS GILL, CO. DURHAM, WHICH OPERATED UNTIL 1958.
The last beehive ovens to operate in Great Britain. Each curved iron 'crane' in front of the ovens supported the handle of a large coke shovel, when the oven was being emptied. A shovel is visible, lying against ovens, between the two nearest cranes.

Fig. 56. BY-PRODUCT RECOVERY OVENS AT CROOK, CO. DURHAM, BUILT IN 1882.
The steam-powered machine was a *ram* to empty the ovens.

Fig. 57. DECORATION ON A MAP OF ABOUT 1750, SHOWING SALTMAKING. Map of County Durham by Thomas Kitchin.

Fig. 58. The Glass Cone at Lemington, near Newcastle, Built in the late Eighteenth Century.

Preserved by Glass Tubes and Components Ltd.

Fig. 59. Steel-making Furnace near Rowlands Gill, Co. Durham, Built about 1750.

Jars, 1775, illustrates one almost identical "near Newcastle".
Reserved for the Northern Regional Open Air Museum.

Fig. 60. PART OF A PRINT OF NEWCASTLE, 1819, SHOWING TWO LOADED KEELS.
By T. M. Richardson. Note the stack of grindstones in the foreground; beyond is a loaded keel and further out a keel with hoisted sail.

Fig. 61. St. Hilda Colliery, South Shields, about 1880.
To the left are the headstock and screens (cf. Fig. 7). Right of engine house is the balance chain (cf. f on Fig. 8).

No. 8 account of work wrought at East Kenton Colliry with the Hewing, Putting and other underground Charges thear on from April the 7 to and with 21 1802 Jos. Fairs over man.

Hewers	Th	Fr	Sa	M	Tu	We	Th	Fr	Sa	M	Tu	We	Round Coalls xx cor	Small Do.	Wet @2d	Rambel @ 2d	Dull @3d	Tope @4d	£	s	d
Jos. Hunter	-	-	-	16	8	20	3	25	24	-	-	-	7..8	0..17	1..0	2..0	2..0	3..0	0	18	6
Jos. Richie	-	-	-	20	23	28	25	25	19	-	-	-	8..10	0..15	1..0	2..0	2..0	3..10	1	0	8
Rob. Walton	18	24	-	23	24	25	25	25	20	-	-	-	11..8	1..2	-	1..10	7..0	4..5	1	8	5
Jn°. Dixon	24	30	-	25	28	29	30	4	22	-	-	-	11..15	0..19	1..0	2..0	5..0	3..5	1	8	2
Patt. Ramsey	29	32	-	25	28	25	25	25	5	-	-	-	12..1	0..16	-	2..0	7..0	2..10	1	2	9
Jn°. Proudlock	12	30	-	-	24	29	28	20	15	-	-	-	9..9	0..17	1..0	2..0	4..0	2..10	1	4	8
Wm. Brown	32	32	-	13	18	22	18	30	24	-	-	-	10..8	1..5	-	2..10	2..10	3..10	1	14	0
Rob. Melven	32	31	-	28	27	30	30	30	24	-	-	-	14..6	1..5	-	2..10	4..10	2..10	1	5	9
M. Maugham	22	-	-	25	25	16	25	26	21	-	-	-	10..10	1..5	1..10	2..10	5..0	3..10	1	6	7
Jn°. Moor	24	24	-	26	21	20	26	25	23	-	-	-	11..4	0..16	2..10	2..10	5..0	1..10	1	5	10
Jn°. Brown	25	15	-	8	37	28	32	26	24	-	-	-	12..1	1..5	2..10	2..0	3..10	5..10	1	10	9
Jn°. Ramsey	18	25	-	27	30	29	30	25	28	-	-	-	13..6	1..4	-	3..10	4..10	2..10	1	11	9
Joseph Hunter	30	32	-	30	25	25	27	29	24	-	-	-	13..10	1..5	4..10	4..0	8..0	1..10	1	12	9
Jn°. Moor	24	36	-	26	25	22	26	25	23	-	-	-	11..18	0..17	3..0	3..0	5..0	3..10	1	7	10
Hen. Bolam	30	36	-	26	25	25	30	33	36	-	-	-	15..0	1..0	-	3..0	5..10	4..10	2	13	4
Thos. Soulsby	27	33	-	32	30	30	31	32	24	-	-	-	14..14	1..6	5..0	3..10	10..0	2..10	2	11	8
Matt. Bates	30	28	-	17	30	30	30	20	26	-	-	-	13..0	0..17	4..0	3..0	4..0	3..5	1	15	6
Jas. Stenson	25	26	-	26	25	26	19	21	30	-	-	-	11..17	1..8	2..10	3..0	5..0	1..15	1	9	3
Jos. Cameron	-	30	-	25	25	4	16	24	13	-	-	-	8..8	0..17	-	-	4..0	5..0	1	8	0
Thos. Wright	30	-	-	21	21	29	13	19	-	-	-	-	9..10	-	-	-	4..0	5..0	0	18	9
Pett. Hopper	20	20	-	18	30	31	32	32	25	-	-	-	13..11	0..19	4..0	2..0	3..10	3..10	1	12	2
Thos. Maugham	10	1	-	20	17	15	23	9	-	-	-	-	7..6	0..19	2..0	1..10	2..10	1..5	0	18	1
Wm. Hunter	-	-	-	-	-	3	8	5	-	-	-	-	0..16	0..3	-	-	-	-	1	1	11
Scoors	23	23	-	23	27	27	26	26	21	-	-	-	250	22	40	46	87	65	30	3	9
Corves	2	15	-	14	3	4	16	15	0	-	-	-	15	16	10	10	0	15			

Putters													R.coals xx cor	S.coals xx cor	Consid. @6d						
Jn°. Peal	74	59	-	77	58	105	68	78	43	-	-	-	60	60	1½				1	15	3
Cris. Crawford	57	86	-	63	100	71	55	80	53	-	-	-	61	61	1½				1	15	6
Matt. Newton	54	60	-	61	87	56	81	71	49	-	-	-	50	106	1				1	14	9
Jno. Newton	89	60	-	20	96	57	67	77	47	-	-	-	98	70	-				1	14	9
Rob. Hunter	88	60	-	78	58	72	67	52	51	-	-	-	56	60	1½				1	13	3
Thos. Kirkby	46	-	-	67	58	62	51	63	83	-	-	-	53	60	-				1	8	9
Pett. Tood	54	-	-	61	86	55	66	70	49	-	-	-	55	60	0½				0	5	0
Thos. Wright	-	80	-	-	-	-	-	-	-	-	-	-	4..0	-	2				0	4	0
Thos. Maugham	-	70	-	-	-	-	-	-	-	-	-	-	3..10	-	1				0	2	10½
Matt. Bates	-	-	-	47	-	-	-	-	-	-	-	-	2..7	-	1½				0	4	0½
Jo. Cameron	-	-	-	-	66	-	-	-	-	-	-	-	3..6	-	5				0	16	7
Jas. Hunter	-	-	-	-	-	81	-	-	-	-	-	-	13..12	-	16½				£ 13	2	6
Scoors	23	23	-	23	27	27	26	26	21	-	-	-	250	3	0						
Corves	2	15	-	14	3	4	16	15	0	-	-	-	0	15	@						

Trapers													Days								
Jno. Walton	1	1	-	1	1	1	1	1	1	-	-	-	10at	8per					0	6	8
Jno. Moor	1	1	-	1	1	1	1	1	1	-	-	-	10at	8per					0	6	8
Jas. Hopper	1	1	-	1	1	1	-	1	1	-	-	-	10at	9per					0	7	6
Drivers													Days	@					£ 1	0	10
Jas. Wilson	1	1	-	1	1	1	-	1	1	-	-	-	10at	18per					0	15	0
Wm. Newton	1	1	-	1	1	1	1	1	1	-	-	-	10at	18per					0	15	0
Pett. Hopper	1	-	-	1	1	-	1	1	1	-	-	-	10at	16per					0	13	4
Jos. Peal	1	1	-	1	1	1	-	1	1	-	-	-	10at	12per					0	10	0
																			£ 2	13	4

To Hewing xx cor 250..40 of Round Coals at 2ˢ Per score ... 25. 0. 0
Hewing 22..15 of small Do at Do stowed Underground xx cor ... 2. 5. 6
To wet working 40..10 at 2ᵈ Per, Rambel 46..10 at 2ᵈ Per xxcor ... 0. 14. 6
To working Doulle Boreds 87..0 at 3ᵈ. Under the Tope Coal 65..15 at 4ᵈ xx cor ... 2. 3. 8
Putting 250..0 at 12ᵈ Per Of Round Coals to tramins at 12ˢ Per scores xx cor ... 12. 10. 0
Putting & stowing 3..15 Small Coals at Do xx cor ... 0. 3. 9
Consideration for Hewers Putting 16½ Days at 6ᵈ Per ... 0. 8. 3
Ladin Coals 2..5 at 2ˢ⁴ Per ... 0. 4. 6
Shoveling tram Road 2 weeks at 12ᵈ Carrying Picks 2 weeks at 12ᵈ ... 0. 4. 0
Measuring small Coals 10 Days at 6ᵈ and taking out Broking waggons 7 at 2ᵈ per ... 0. 6. 2
attending teamins 19 Days at 2ˢ..6ᵈ per and 10 Do at 16 per ... 3. 0. 10
Runing waggons 3½ Days at 12ᵈ Per ... 0. 3. 6
Keeping trap Doores 20 Days at 8ᵈ Per and 10 Do at 9ᵈ Per ... 1. 0. 10
Driving waggon Horses in 20 fathom Of Do 10 Days at 18ᵈ Per and 20 Do at 14ᵈ per ... 2. 13. 4
setting on the waggons at Bottom of Do 10 Days at 3ˢ..10ᵈ Per and 10 Do at 3ˢ..4ᵈ Per ... 2. 11. 8
Banking out at Top. of Do 10 Days at 2ˢ..6ᵈ Per and 10 Do at 2ˢ..2ᵈ Per ... 2. 6. 8
attending Machine at Tope of Do 2 Men each 1½ Weeks at 15ˢ Per ... 2. 5. 0
Setting on at Kenton Pit and attending the Furne. 12 Days at 2ˢ..4ᵈ Per ... 1. 8. 0
Banking out at Kenton Pit 2 Weeks at 16 overtime 4ˢ Per ... 1. 16. 0
Attending Machine at Tope of Do. 2 at 15ˢ Per Lodgings 2ˢ. Owed 3ˢ ... 1. 15. 0
Putting throu' troubles in the Bordes 2ˢ Per ... 0. 4. 0
Lodgings for 2 men Each 2 weeks at 12ˢ Per ... 0. 10. 0
Chearity Muney 4 weeks at 2ˢ..6ᵈ Per ... 0. 5. 6
To putting and keeping Trap Doors 5½ Days at 12ᵈ per ... 9. 7
Strowing Propps and Plancks 230 at ½ᵈ per ... 7. 5. 10

setting on the Pit Holling walls finding Deputes and propp trailers, xx cor 250..0 at 7ᵈ Per
Candels for Putters and Drivers

£ 71. 18. 1

Exᵈ TE

Deductions
To 25 Corves laid out at 3ˢ per ... £0. 6. 3
To 60 do Setout at 2ˢ per Score ... 0. 6. 0
To 60 do do putting at 12½ᵈ per ... 0. 3. 0
0. 15. 3
£ 71. 2. 10

Fig. 62. An Account of Work. 1802 (see page 47).

Saltmaking

Salt was being made on the south coast at the time of the Domesday Book, by burning wood to evaporate sea-water. As coal began to be mined in the north we find references to its use for saltmaking on the north-east coast. Early mention of saltmaking there cannot be proved to refer to the use of coal, but in 1463 the monks of Tynemouth obtained a charter from Edward IV for loading and unloading their own and other ships, to enable them to trade freely in coals and white salt. An early reference to 'iron' salt pans at South Shields occurs in 1499, though lead continued in use as the chief material for some time after.[1]

In 1635 Sir William Brereton travelled through the north of England and in June he went to "Tine-mouth" and the "Sheeldes": "Here I viewed the salt works, wherein is more salt works and more salt made, than in any part of England that I know, and all the salt here is made of salt-water". Sir William then went on to describe these works and the pans in detail.[2] Ninety years later Lord Harley, whilst travelling north, visited South Shields: "which is the chief place for making salt. The houses there are poor little low hovels, and are in a perpetual thick nasty smoke. It has in all 200 salt pans, each employs 3 men. Each pan makes one tun and a $\frac{1}{4}$ of salt at 8 boilings which lasts 3 days and a half. Each consumes 14 chaldrons of coal in 7 days in which time it makes two tuns and a $\frac{1}{2}$ of salt. The wages for Pumpers, *i.e.* those who pump the salt water into the pans is 5d. per diem. The watchers, *i.e.* those who continually have an eye to the pans and the fire stoves have 6d. a day".[3] A "Map of County Durham" by Thomas Kitchin of *c.* 1750 incorporates an embellishment showing a corner of a salt pan and a salt-shovel (Fig. 57). In addition to this illustration, Kitchin added: "South Shields, the Station of the Sea Coal Fleets, is a large Village eminent for its Salt Works, here being upwards of 200 Pans for boiling the Sea Water into Salt. 'Tis said that 100,000 Chaldron of Coals are yearly consum'd in these Works".

Equipment for saltmaking, identical with that illustrated by Thomas Kitchin, may be seen in greater detail in William Brownrigg's book *The Art of Making Common Salt* (1748). Brine was pumped from the sea at full tide into a brine pit and run from there by lead pipes to each pan. The boiling pans measured about 20 ft by 12 ft by 14 in. deep, built of plates of iron rivetted together with iron nails and the joints filled with a cement. Across the top of the pan were placed several strong bars of iron and, from these, iron hooks hung down to support the pan bottom at intervals. Thus the large area of pan bottom was kept in shape. (An account of 1722 from the Cotesworth MSS. details the cost of building a new 'Pann and Pannhouse' at £145 13s. 7$\frac{1}{2}$d. of which £86 8s. 10$\frac{1}{2}$d. was for iron plates and bars.)[4] A series of grates beneath the pans boiled the brine and as this evaporated, salt was skimmed and shovelled off and packed into wicker baskets to drain. Three of these conical baskets can be seen in Kitchin's illustration.

Despite this eighteenth century success, the salt trade of the north-east declined and at the time that Surtees was writing his *History of Durham* (1820), he recorded that only five salt pans remained at South Shields.

[1] Surtees, 1820: II. 94.   [2] *Journey through Durham and Northumberland*, in Richardson, 1849.
[3] Lord Harley: *Journeys in England*, 1725. Hist. Man. Comm. Rep. on MSS. of Duke of Portland, VI, p. 105.
[4] Gateshead Public Library.

Iron and Steel and Glass

By the middle of the sixteenth century the depletion of the country's timber resources threatened a fuel famine which not only overcame the deep-seated prejudice against the use of coal for domestic heating, but stimulated new methods in industry, for example, in glass making. Other industries were directly stimulated by this expanding coal trade. Cheap coal and cheap water-transport were of paramount importance, for small coal that was unsuitable for shipment to London was available locally at an almost nominal price, and bulky raw materials could be cheaply brought to the Tyne since they served as ballast for returning colliers. Such industries as the making of salt, glass, pottery, and copperas developed rapidly (copperas, used as a mordant, was made from iron pyrites found associated with coal). Moreover, the large demand for ships and their accessories, created by the coal trade, provided yet another local industry and shipbuilding expanded with coal shipments. It was the demand for ironwork for ships—nails, anchors, and chain—that brought Ambrose Crowley to the north-east in about 1682, when he set up his works in Sunderland. In 1691 he moved his factory to Winlaton and subsequently began another at Swalwell. Here he and his successors thrived until the mid-nineteenth century. The last chainmaker ceased working in Winlaton in 1966, and a few semi-derelict chain-shops may still be seen, but the industry had effectively ceased by about 1850.

This intimate connection between the coal trade and industrial development explains the essential features of early industrial settlement in north-east England. As Smailes points out, the effective beginnings of this development date from the middle of the sixteenth century, and specialised and distinctively industrial settlements first appeared at that time.[1]

Two remarkable buildings remain to remind us of this early industrial growth. One on the north side of the Tyne can be seen from afar: the glass cone at Lemington, dating from the late eighteenth century. It is probably the largest remaining brick-built glass cone, although they were once common. Within this structure the pots of 'metal' (molten glass) were heated and the glass was blown. It is now preserved by G.E.C. Ltd, and is the last visible link with the Tyneside glass industry, which had its main period of growth and decay between the 1690s and 1790s, with a few firms lasting into the nineteenth century.

Another industrial building which takes one back to the eighteenth century is the Derwentcote steel furnace near Rowlands Gill. It was built for the conversion of iron into steel in the mid-eighteenth century. At that time steel was made by heating iron over a long period in the presence of charcoal: the red-hot iron absorbed carbon and became steel. This could be done by lengthy heating in charcoal fires, but could not be done in coal fires because impurities such as sulphur were also absorbed, making the metal brittle. It was found that if the iron bars were packed in charcoal, in sealed earthenware *chests*, these could be then heated by coal fires without such ill-effects. When Gabriel Jars visited England in 1764 he spent some time in the Newcastle area, and described and illustrated a steel furnace which he saw "near there". This closely resembled the one still remaining at Rowlands Gill.

[1] Smailes, 1960: 131.

Bibliography

Useful modern books

ASHTON, T. S., and SYKES, J.	1929	*The Coal Industry of the Eighteenth Century.*
HUGHES, Edward	1952	*North Country Life in the Eighteenth Century.*
MOTT, R. A. (Ed.)	1936	*The History of Coke-Making.*
NEF, J. U.	1932	*The Rise of the British Coal Industry.*
ROLT, L. T. C.	1963	*Thomas Newcomen.*
SMAILES, A. E.	1960	*North England.*
TOMALIN, M.	1960	*Coal Mines and Miners.*

Recent papers

ATKINSON, Frank	1960	"The Horse as a Source of Rotary Power." *Trans. Newcomen Soc.*, XXXIII, 31-55.
,, ,,	1965	"Some Northumberland Collieries in 1724" (Journal of Sir John Clerk). *Trans. Arch. and Ant. Soc. of Durham and Northumberland*, Vol. XI, 425-434.
ATKINSON, P.	1956	"The Isabella Winding Engine." *Journ. Stephenson Engineering Soc.*, I. No. 4, 75-97.
BUCKLEY, F.	1926	"Glasshouses on the Tyne in the Eighteenth Century." *Trans. Soc. Glass Technology*, Vol. 10.
LAW, R. J.	1965	"The Steam Engine", *A Science Museum Booklet*, London.
LEE, C. E.	1951	"The Wagonways of Tyneside", *Archaeologia Aeliana*, Vol. XXIX, 135-202.
,, ,,	1960	"Some Railway Facts and Fallacies", *Trans. Newcomen Soc.*, XXXIII, 1-16.
SMITH, R. S.	1957	"Huntingdon Beaumont", *Renaissance and Modern Studies*, I, 115-53.
WATKINS, G. M.	1955	"Vertical Winding Engines of Durham." *Trans. Newcomen Soc.*, Vol. XXIX, 205-19.

Books used for illustrations and quotations

AGRICOLA, Georgius	1556	*De Re Metallica.* Trans. and ed. by H. S. and L. H. Hoover, London, 1912 and New York, 1950.
BAILEY, John	1810	*Agriculture of County Durham.* (Report to the Board of Agriculture.)
BALD, Robert	1812	*Coal Trade of Scotland.*
BELL, J. T. W.	1843-61	*Plans of the Great Northern Coalfield.*
BOURNE, Henry	1736	*History of Newcastle Upon Tyne.*
BRAND, John	1789	*History and Antiquities of Newcastle.*
BROWNRIGG, W.	1748	*The Art of Making Common Salt.*
BUDDLE, John	1814	*First Report on Accidents in Coal Mines.*
C., J.	1708	*The Compleat Collier.*
CURR, John	1797	*Coal Viewer and Engine Builder's Companion.*
DEFOE, Daniel	1726	*The Complete English Tradesman.*
DUNN, Matthias	1844	*View of the Coal Trade.*
	1848	*Winning and Working Collieries.*
EDINGTON, Robert	1813	*Treatise on the Coal Trade.*
EMERSON, William	1758	*Principles of Mechanics.*
GALLOWAY, R. L.	1898	*Annals of Coal Mining and the Coal Trade.*
GIBSON, J.	1787	*Plan of the Collieries of the Rivers Tyne and Wear.*
GREENWELL, G. C.	1855	*Practical Treatise on Mine Engineering.*
,, ,,	1888	*Glossary of Terms used in the Coal Trade of Northumberland and Durham.*
HAIR, T. H.	1844	*Coal Mines of Durham and Northumberland.*
HODGSON, Rev. John	1813	*Funeral Sermon: Felling Explosion.*
HOWITT, W.	1841	*Visits to Remarkable Places.*
HOLLAND, John	1835	*Fossil Fuel.*
HUGHES, H. W.	1892	*Text Book of Coal-Mining.*
JARS, Gabriel	1774 & 1781	*Voyages Metallurgiques.*
MORAND, J. F. C.	1768-77	*L'Art d'exploiter les Mines de Charbon de Terre.*
NORTH, Roger	1742	*The Life of Lord Guildford.*

PLOT, Robert	1686	*Natural History of Staffordshire.*
PRIMATT, Stephen (2nd ed.)	1680	*The City and County Purchaser and Builder.*
RICHARDSON, M. A.	1849	*Reprints of Rare Tracts of Northern Counties.*
SINCLAIR, George	1672	*The Hydrostaticks.*
STUKELEY, W.	1776	*Itinerarium Curiosum.* Cent. II.
SURTEES, Robert	1816-1840	*History of Durham.*
TOMLINSON, Charles	1854	*Cyclopaedia of Useful Arts and Manufactures.*
TOMLINSON, W. W.	1914	*The North Eastern Railway.*
VICTORIA COUNTY HISTORY OF COUNTY DURHAM	1905, 1907, 1928	
WALLIS, John	1769	*Natural History and Antiquities of Northumberland and Durham.*

Index

ACCIDENTS, 21, 30, 42-3

'BANK', 'banksman', 37
Bank Top Station, 55
Beamish Colliery, 34, 35
Beaumont, Huntingdon, 6
'Bell-pit', 6
Benwell Estate, 23
Bewick, John, 31
— Thomas, 31
'Biggins', 11
Birtley Colliery, 52
Blasting, 9
Blyth Colliery, 51
Bord and pillar, 11, 12, 13
— Dimensions, 11
Boring, 6-7
— (chisel-bit), 6-7
— (hollow rods), 6-7
Bracke, 7
Brand, John, 6-7, 27
Brattices, 9, 10, 42
Brereton, Sir William, 70
Brockwell seam, 12, 13, 14
Brusselton Incline, 55
Buddle, John, 21, 42
Burradon Colliery, 42
Butterknowle, 11
Byker, 18
— Colliery, 26

CARTRIDGES, 16
Chaldron wagons, 38, 49, 50
Chance Pit, 62
Charcoal, 62
Chester-le-Street, 37, 52
Chiffonier, 44
'Cinders', 62
Clanny, Dr William, 21
Claughton, George, 62
Clerk, Sir John, 18, 23, 27, 31, 51
Coal drops, 57, 59, 60
Coal-picking, 38
Coal screening, 37, 41
Coal-washing, 38
Coke and coking, 62
Coking, by-products of, 63, 64

Consett, 55
Copperas, 71
Corf, corves, 5, 27, 29, 30
Coxhoe, 44
'Cracket' (stool), 14
Crawleyside, 55
Crook, 63, 64
Crowley, Ambrose, 71
Crowther, Phineas, 35
Curr, John, 30, 35, 52

DAVY, Sir Humphrey, 21, 42
Defoe, Daniel, 57
Delaval Drift, 23
Delavel, Sir Ralph, 26
Derbyshire, 5, 30
'Dess-beds', 44
Double bord, 13, 47
— working, 13
Drainage, 23-6
— by buckets ('tub'), 23
— — engine, 26
— — rag and chain, 23
— — watergates, 23
Drilling, 14
— by 'monkey', 14
— — 'puncher' or 'jumper', 14
Dunston, 55
Durham, 5

EAST Kenton Colliery, 47, 69
Elemore Colliery, 35, 42
Eltringham, 31
Engine, atmospheric, 26, 27
— steam, 27, 35
Erecting pitmen's houses, cost of, 44
Etherley Incline, 55

FATFIELD Colliery, 21, 42
Felling Colliery, 12, 21, 30, 42
Finchale, 23
'Flat', 30
'Foal', 30

'GARLANDS', 9
Gas, 'choke' damp, 16
— dealing with, 11, 18
— explosive, 42
Gaunless, River, 55
Gin, cog and rung, 31, 32, 34
— whim, 31, 32, 35, 36
Glass, 66, 71
Gosforth Colliery, 62
Grimm, 44
Gunpowder in mining, 9, 13

HARLEY, Lord, 70
Hartley Colliery, 42
Haswell Colliery, 42
Haulage by inclined plane, 30
— steam, 52
'Headsman', 30
Hebburn Colliery, 17, 37
Hett, 23
Hewing coal, 13-16
Hodgson, Rev. John, 42
Housing of pitmen, 44-5
Hownes Gill, 55
'Hoys' (colliers), 61
Hunt, Thomas, 35
'Hurrying' coal, 5

IRON working, 71

JARS, Gabriel, 27, 62, 71
'Jenkins', 11

'KEELS', 57, 61, 67
Kell, John, 12, 13
'Kerving' out, 13, 14
Kilkenny, 5
Killingworth Moor Colliery, 35
Kitchin, Thomas, 70

LANCHESTER Moor, 9
Leasingthorne Colliery, 12, 14
Leifchild, J. R., 47
Lemington, 66, 71
Liége, 16

Lighting, 21-2
— by candles, 21
— — Geordie lamp, 21
— — midgie lamp, 21
— — phosphorus, 21
— — putrescent fish, 21
— — steel mill, 21
Locomotive railways, 55
'Long' and 'short' (methods of mining), 11

'MARRERS', 13
Meeting Slacks engine, 55
Moray, Sir Robert, 16
Morpeth, 51

NEWBIGGIN, 23
Newbottle Colliery, 42
Newcastle, 6, 23, 26, 30, 31, 44, 57, 62, 67
Newcomen, 24, 26
North, Roger, 48
Northumberland, 5

OVENS, beehive, 62, 63, 64
Oxclose Colliery, 26

'PANEL working', 11
Pelton Colliery, 42
— Incline, 53
Piece work payment, 27
Pillar and stall, 11
Pitmen's clothing, 46, 47
— furniture, 44
Pontop Colliery, 48
Potter, John, 27
Pottery, 71
Prices and piecework rates, 13, 47
Primatt, Stephen, 23
Primitive Methodists, 44
'Putters', 5, 27, 30, 47

RAILS, 48, 51
Railways, 30
— Brandling Junction, 55
— Stanhope and Tyne, 55, 56
— Stockton and Darlington, 55
Rents paid for housing, 44
Ridley, Richard, 18
Rods (boring), 6
'Rolleys', 30
Rowlands Gill, 63, 64, 66, 71
Ryton Colliery, 20

SALT-MAKING, 37, 65, 70, 71
Screening coal, 37-41
Seaham Colliery, 21, 42
— Harbour, 51, 57, 60
'Sheths', 30
Shildon, 55
Shipping, 61
Shot-firing, 9, 10, 13, 14-16
Sinclair, George, 6
Sinking, 8-10
'Soams', 30
South Moor, 48
— Shields, 55, 68, 70
'Spouts', 57, 61
Springy pole, 7
Staithes, 51, 57-8, 62
'Standage', 10
Steam engine, 35
— haulage, 52
Stephenson, George, 21, 51, 52, 55
'Stoppings', 18
Stukeley, William, 61
Swalwell, 71

TANFIELD Arch, 48, 49
— Colliery, 48
— Lea Colliery, 26
— Wagonway, 55
Tipperary, 5
'Tram', 27, 29, 30

Tramways, 5
Transport by corves, 27, 30, 31
— — engine underground, 31
— — horses underground, 27, 30, 31
— underground, 27-31
'Trappers', 18, 47
Trimdon Grange, 42
'Tubbing', 9

VENTILATION, 9, 10, 16-20
— by fire, 16-18, 20

WADDLE fan, 20
Wages, 47
Wagonways, 30
— horse, 48, 50, 52
Walker Colliery, 27
Wallsend, 57
— Colliery, 42
Warden Law, 52, 54
Watt, James, 27
'Wayleaves', 48
Weatherhill, 55
Wesleyans, 44
West Kenton, 23
Whim gin, 31-2 35, 36
Whorlton, 23
Winding, 31-6
— by cog and rung gin, 31
— — hand windlass, 31
— — horse, 31
Winlaton, 71
'Winnin, the', 12
Witton Park Colliery, 55
Wood, Ralph, 48

YORKSHIRE, 5, 11, 13, 30